The book Racing Car Builder has been planned in an effort to aid the novice car builder in the construction of a racing car. The instructions have been set forth in a manner which makes them applicable to the construction of either a midget racing car or a three-quarter (dirt track) car. Highly technical terminology and engineering phrases have been deliberately avoided, the aim being rather to make the book clear and understandable to all, regardless of technical background or training. The fundamental principles described in this book may also be applied to the construction of a track roadster or "hot rod", though the bodies of these cars differ from those of standard racing cars.

The first requisites which determine the dimensions and type of construction of a car are the rules and regulations of the governing body or association in which the car is going to operate. Obviously the car must conform to the regulations of the operating association. For this reason then, the first section of Racing Car Builder sets forth the approved regulations of the three largest operating organizations in the United States, the American Automobile Association Contest Board, the Central States Racing Association, and the United Racing Association. The AAA operates nationally, the CSRA in the midwestern states and the URA in the Southwest. In addition, in an attempt to show the car builder the stringent rules which the builders of Indianapolis Championship cars must adhere to, we have reprinted section 5 of the supplementary regulations of National Championship races of the AAA Contest Board. These latter rules do not apply to midget or dirt track cars, but are shown only to give the reader an idea of the many things which the builders of Indianapolis cars must consider during construction of their entries.

It is sincerely hoped that the information contained herein will benefit the reader in whatever phase of racing car building he is undertaking...........Floyd Clymer

Copyright 1949

Published by

FLOYD CLYMER

America's Largest Publisher of Books Pertaining to
Automobiles, Motorcycles and Racing

1268 So. Alvarado St. • Los Angeles 6, Calif.

HOW TO BUILD A RACING CAR
By K. J. Domark

TABLE OF CONTENTS

ILLUSTRATIONS

APPROVED SUPPLEMENTARY REGULATIONS
NON-CHAMPIONSHIP RACES

The following additional rules and conditions have been approved by the Contest Board of the American Automobile A ciation, and govern the conduct of the meeting with the same force as the Official Competition rules. They will be str enforced. The management wants to treat all entrants alike, believing that it is only good sportsmanship to do so.

1. **ENTRY**—Shall consist of a combination of car and driver. Entry blank must be completely filled out, including the duplicate and complete data given on every question. ENTRIES POSITIVELY CLOSE at midnight of the date specified unless stated to the contrary on the title page. Entries may be made by telegraph. False statements will automatically void an entry. See rule book.

2. **ELIGIBILITY**—All drivers and cars must be registered by the Contest Board, A.A.A. Entries may be made by non-registered cars and drivers. If you are not registered, in filling out this entry blank so indicate in the space provided for car and driver registry number. The registration will be considered by the Stewards at the drivers' and officials' meeting preceding the race. See rule book.

3. **FEES FOR REGISTRATION**—Drivers' annual license, Jan. 1 to Dec. 31, $15.00. Temporary Permit, $4.00. Annual racing car license, Jan. 1 to Dec. 31, $10.00. Temporary Permit, $4.00. Mechanics' annual license, Jan. 1 to Dec. 31, $5.00. Temporary Permit, $2.00. Permits may be issued only to novice drivers and cars of unknown origin.

4. **CANCELING EVENTS**—The management reserves the right to cancel the event unless a satisfactory number of entries have been received by the date on which entries close. In the event of cancellation, notice will be made to all entries by wire or postal card.

5. **CAR NUMBERS**—The assignment of car numbers will be made by the Contest Board or Representative. Car numbers must be prominently painted as large as practical in legible design on all cars on both sides of the hood and left side of fuselage. Numbers such as 30X and A1, numbers over 99, numbers 11 and 13, ARE NOT PERMITTED to be used. Each car will carry its name on each side in letters not over 8 inches in height. If the car already carries a number, so indicate, when making your entry. In cases of duplication of numbers see rule book.

6. **EXAMINATION**—All cars may be examined by the Technical Committee as to engine measurement preceding or after the race. Quality of painting and of workmanship will be taken into consideration in determining the degree of preparation that has been made for the contest. As a general rule cars which are makeshift in appearance are makeshift mechanically and vice versa and such cars will be barred from competition.

7. **MECHANICAL REGULATIONS**—All cars must be equipped with a suitable braking device as required for the type of event in which it is to participate. Steering wheel spiders must be either steel or bronze. Steering wheel mechanism, knuckles and spindles must be approved by the A.A.A. Representative or Technical Judge. Exhaust pipes must be extended beyond rear of driver's seat. Metal fire dash is compulsory. Carburetor fittings of brass are required by the Technical Committee if, in its opinion, die cast parts are insufficiently supported or reinforced. Fuel system must be equipped with a shut-off device placed within easy reach of the driver. Mufflers may be required on any or all cars when, in the opinion of the Stewards, local conditions necessitate their use. Each car is required to be equipped with a declutching device. Cars must be equipped with an ignition switch or "shorter-button" within easy reach of the driver. A suitable fuselage or tail will be compulsory and must extend beyond the rear wheels. A full hood is required, which must be securely fastened to the body or frame of the car, to the satisfaction of the Technical Committee. Gasoline Tank must be properly constructed with all joints lapped, riveted and sweated, or lapped and welded. Tanks that are not properly supported or are unduly exposed may be rejected by the Technical Committee. A minimum ground or road clearance of four inches will be required. Cars will be required to carry a metal underpan extending from the front of the driver's seat to a point at least six inches forward of any foot-control or foot-rest, and which must extend from side-member to side-member of the frame for the full length of the pan. Vent holes in the under-pan should be located so as to prevent leakage of fuel, oil or water on to the track. All rotating parts within the cock-pit must be protected by a suitable guard. It may be required by local A.A. Contest Board officials that the radiator overflow pipe be terminated in a (vented) overflow tank to prevent water dripping on to the track. This tank must be equipped with a drain cock for emptying.

8. **STARTING**—Rolling starts will prevail unless otherwise stated. Cars will be started as specified on the title page or on supplementary regulations. In case of two or more cars tying in qualifying, the car to qualify first shall have preference. See rule book.

9. **CARS TO START**—The surface and the condition of the track are the controlling factors. This entry blank indicates on its face the maximum number of starters allowed. The Representative of the Contest Board in attendance, however, may rule, after consultation with the Stewards, that the field should be further restricted where proper regard for safety so suggests. In the event of more than the specified number of starters being entered, the fastest cars in the official speed trials up to and including that number will be eligible; provided, however, that the driver's ability and measurement and design of car be approved by the proper officials.

10. **QUALIFICATION TRIALS**—The date and hour of qualification trials will be as stated on the title page or announced to each entry by bulletin signed either by the management or Contest Board Representative. Drivers failing to qualify within the official hours lose position to which trials might entitle them, but may be assigned a last position if the events permit. Entrants on the day's program shall, having run their qualifying trials, report with their cars at the pit line or paddock not later than 15 minutes before the hour at which the first event is scheduled to start. See rule book.

11. **HEATS**—When more than the maximum number of starters are e and especially where the day's program consists of a number of events, the management reserves the right to hold heat or elimi trials other than specified under schedule of events; distance and tions to be approved by the Stewards.

12. **DISQUALIFICATION**—No entrant shall have any claim for dar expense, or otherwise against the management or any of the offici Representatives by reason of disqualification of either car or driver

13. **APPEARANCE**—All drivers, car owners and pit crews must be c in clean appropriate uniforms. Carelessness will not be tolerated. people in the grandstand are entitled to see clean, well-dressed formers. It is their money that makes professional racing po Grease-soaked clothing will not be tolerated.

14. **DISCIPLINE**—It is expected that the deportment of everyone con will be that of a gentleman and sportsman at all times, those not co ing themselves in this manner will be subject to discipline.

15. **MINIMUM AGE**—NO PERSON UNDER 21 YEARS OF AGE WIL ELIGIBLE AS DRIVER, MECHANICIAN OR PIT ATTENDANT.

16. **COMPETENCY OF DRIVERS**—Any driver who, in the opinion Stewards, does not show sufficient skill and judgment in the handli his car to make him a safe factor in competition will be declared ine to participate in these events or the practice period incident thereto.

17. **PHYSICAL CONDITION OF DRIVERS**—Any driver who, on th of the race, gives evidence of exhaustion, drink, or other physical pacity, making him a potential danger to others on the course, m barred from further competition, at the discretion of the Stewards.

18. **RELIEF DRIVERS**—No driver will be accepted as a relief or subs driver unless he has driven the car, or one of the same kind and s during the practice period.

19. **EXHAUST**—Cars liberating an excessive amount of oil or smoke be disqualified by the officials as dangerous to other contestants.

20. **ANY CAR** developing a gasoline leak during the progress of an will be required by the Starter or Stewards to stop at the pits o succeeding lap for examination; failure to observe this rule ma considered by the Stewards as cause to disqualify the entry for remainder of the program.

21. **PIT RULES**—Each entrant will be allowed necessary pit tags a seats. These will be distributed by the Stewards. The following lations as to conduct will govern at every meet held under A sanction. Immediate responsibility for their observance is attache the Technical Judge or Committee. The Stewards, however, are ch with making necessary arrangements, including a police detail, to s the results.

   A. Proper deportment in each individual pit is charged to the en (if present) or driver.

   B. Pitmen must remain as far as possible in their own pit. Promise visiting will not be permitted.

   C. No one will be allowed on the track or in the pits without the p tag, which must be visible at all times.

   D. There must be no standing or sitting on any rails. (It inter with the view of the field.)

   E. Cars must be kept arranged in an orderly manner parallel to the tr if possible, and close to the pit.

   F. All mechanicians and pit attendants must be registered.

   G. Pitmen must keep their eyes on their own car at all times, read help their driver. That is what they are there for and nothing

   H. Neatness and cleanliness are required in pitmen as well as in dr GREASE-SOAKED CLOTHING WILL NOT BE TOLERATED

Unless otherwise covered in the Supplementary Regulations prem will issue pit passes only as follows:

In major events, each entrant will be allowed a staff of five men for car—a mechanician, and four general assistants—and in case of a for each additional car, three men—a mechanician, and two assist One of the group may be named team manager. Mechanicians wil furnished distinctive credentials from pitmen. In local events, a sta three men to a car is considered adequate (driver, mechanician helper). In no case shall there be allowed more than five.

22. **PAYMENT OF PRIZE MONEY**—Prize money will be paid to TRANTS of winning cars under A.A.A. supervision twenty-four h after the finish of the race meeting, except in case of protest. See book. The A.A.A. Representative may withhold the prize money by any car until the Technical Committee has inspected and app the car as complying with the supplementary regulations. No fo protest is necessary to instigate this inspection. Cars whose mea ments do not comply with the limits specified in these regulations ineligible for competition under these rules. Any prize money by such an ineligible car will, upon proof of non-compliance, be decl forfeited.

MOTOR AND CHASSIS SPECIFICATIONS

Small Car or Midget Class

ON DISPLACEMENT—Shall be limited to a maximum of 105 cu. in. on engines employing overhead valves or engines which are so designed that the major portion of any valve seat is located above a plane normal to the bore and coincident with the uppermost level reached by the full diameter of the piston when it is at top dead center. A maximum displacement limit of 140 cu. in. will be allowed on engines using a valve-in-block design where all valve seats are below the junction of the cylinder head with the cylinder block. Superchargers will be permitted on engines of 52 cu. in. displacement or less.

CHASSIS SPECIFICATIONS—Wheel base limits will be restricted to a minimum of 66 inches. Tread width will be restricted to a maximum of 46 inches and a minimum of 42 inches, measured between center-lines of opposite tires at point of contact with the track. The diameter of rims will be limited to a minimum of 12 inches. Rims must be equipped with suitable driving lugs or pins. Wheel design and construction will be subject to approval of the Technical Committee. In general there will be no restrictions on the maximum tire cross section. However, such restrictions may be approved by the Contest Board for special events or certain tracks where special or unusual conditions exist.

Regular Class

ON DISPLACEMENT—Shall be limited to a maximum of 220 cu. in. on engines employing overhead valves or engines which are so designed that the major portion of any valve seat is located above a plane normal to the bore and coincident with the uppermost level reached by the full diameter of the piston when it is at top dead center. A maximum displacement limit of 274 cu. in. will be allowed on L-Head and Flat-Head engines. Superchargers will be permitted on engines of 138 cu. inches displacement or less.

CHASSIS SPECIFICATIONS—Wheel base limits will be restricted to a minimum of 84 inches.

General Regulations

No driver will be permitted on the track for practice or competition without wearing an approved crash helmet and shatter-proof goggles.

Driving on the inside shoulder of a track or in the infield during or in connection with a race or practice period will not be permitted.

During a competition passing may be done on either side with due regard to safety.

Flag signals will conform to the colors and usage prescribed by the International Sporting Code and the Official Rules of the A.A.A. Contest Board. The following flag signals will be enforced:

GREEN FLAG or LIGHT—Start; course is clear.

YELLOW FLAG or LIGHT—Caution; watch for conditions ahead; get your car under control, and HOLD YOUR POSITION.

RED FLAG or LIGHT—Stop immediately; race is halted.

KING BLUE WITH DIAGONAL YELLOW STRIPE—Passing signal; faster car is endeavoring to pass; blocking prohibited.

BLACK FLAG—Stop at pits next lap for instructions.

WHITE FLAG—Entering last lap.

CHECKER FLAG—You have finished race.

Every participant is expected to secure a copy of, and become familiar with the Official Competition Rules. The above is general guidance ONLY, and ignorance of the Official Rules will not be accepted as a plausible excuse for any infraction.

5. **ELIGIBILITY**—

(a) DISPLACEMENT—Supercharged engines will be limited to a maximum piston displacement of 183.060 cubic inches (3,000 cc.). Non-supercharged engines will be limited to a maximum piston displacement of 274.59 cubic inches piston displacement (4,500 cc.).

(b) WHEELBASE—The minimum wheelbase shall be ninety-six (96) inches. The A.A.A. Technical Committee shall have the right and power to determine the safety factors and handleability of all cars with respect to this regulation.

(c) WEIGHT—(No limitations).

(d) BODY TYPE—The body may have one or two seats.

(e) CARBURETOR fittings of brass may be required by the American Automobile Association Technical Committee if in its opinion die cast parts are insufficiently supported or reinforced.

(f) BRAKES—Each car must be equipped with a braking system which must operate the brakes effectively on all four wheels.

(g) TRANSMISSION—The transmission system must incorporate a declutching device and a reverse as well as a forward speed.

(h) FUEL TANK—Fuel tank must be constructed and tank supported in such a manner as to insure against breakage. The construction and suspension of fuel tank and fuel lines shall be subject to the approval of the American Automobile Association Technical Committee.

(i) ENGINE LUBRICATING OIL SUPPLY—Oil supply for entire engine lubricating system unlimited. Engine oil supplies may be allowed in the pits after race starts at the discretion of the Stewards. Loose containers of oil may not be carried in the cars. Note Paragraph (o), Section 3, which will be strictly enforced.

(j) STARTING ENGINES AT PITS—Pushing cars at pits to start engines will not be permitted. All engines must be started at pits by self-starter or hand crank.

(k) FIRE WALL—A fire wall is required between the engine compartment and the driver's cockpit. The fire wall construction shall be subject to the approval of the A.A.A. Technical Committee.

(l) STEERING—Steering wheel must have a steel insert for unit rim and spider. Steering gear ratio may not exceed 15:1 unless approved by A.A.A. Technical Committee.

(m) TIRE RING LINK LOCKS—Wheels may be required to be equipped with an approved tire ring link lock. Balancing lugs for wheels and tires must be approved by the A.A.A. Technical Committee. No lugs of more than two inches in length will be permitted and lugs must be securely fastened.

(n) PARTS—American Automobile Association Technical Committee may require that steering mechanism or any other parts be new and proportioned to the car. The Chairman of the Technical Committee may demand or permit any changes pertaining to safety. The American Automobile Association shall have the right to require Magnaflux examinations of crankshafts and/or other parts. However, such examination may be made and certified to before arrival at Speedway, provided certification report of such Magnaflux examination is made by responsible persons or organizations, and states exactly the parts examined and adequately identifies each tested part by distinctive die mark or other suitable methods.

(o) OIL AND GREASE LEAKAGE—Leakage of lubricants will not be tolerated. American Automobile Association Technical Committee may bar any car having excessive leakage from using the track.

(p) INSPECTION AFTER RACE—The management of the race and A.A.A. Contest Board reserve the right to inspect, for non-compliance with the rules, any or all cars finishing the race.

Engine . . . .

1. PISTON DISPLACEMENT: Limited to 140 cu. in. for flat heads, 105 cu. in. for OVERHEAD cams and ROCKER ARM 4 cycle engines. 75 cu. in. for 2 cycle engines. Slide valve engines will be classed as Rocker Arm type. OVERSIZE on an engine will be .009 cu. in. over the above limits.

2. ENGINE DESIGN: Any engine, other than described above, must be approved by the Contest Committee.

3. SUPERCHARGERS: They will be allowed on motors up to 65 C. I.

4. EXHAUST PIPE: Each car must have an exhaust pipe or pipes that extend beyond the rear axle or the cockpit.

5. CARBURETORS: Any type or amount permitted.

6. FOUR WHEEL DRIVES ARE NOT PERMITTED.

Chassis . . . .

1. WEIGHT: Minimum 550 lbs., Maximum 950 lbs., minus gas, oil, water or the driver.

2. WHEELBASE & TREAD: Wheelbase, Minimum 66 inches, maximum 76 inches. Tread, minimum 42 inches, maximum 46 inches. (NOTE) Tread is measured from mounted tire center to center when car is on the track ready to run.

3. HUBS: Hubs must not extend to a danger point and must be *Safety Hubs.* Referee may bar cars from competition until so equipped. No CAST IRON hubs will be permitted.

4. TIRES: No tires with driving studs will be allowed. (Bolts, nails, etc.) No dual wheels or tires allowed.

5. WHEELS: Maximum diameter, 12 inches ONLY. Maximum width 5½ inches. All rims must be equipped with suitable driving lugs or pins and must meet with the approval of the Contest Committee.

6. STEERING WHEELS: All cars must be equipped with flexible steel spider type steering wheel. The steering wheel must be placed so that the driver has easy entrance and exit to the cockpit and be able to handle the car without interference of the wheel.

8. AXLE: Front and rear axle assemblies must meet with the approval of the Technical committee.

9. STEERING ASSEMBLIES: Drag Link, Steering Arms and Spindles must pass the inspection of the Technical Board. Brazing will not be tolerated.

10. FIRE WALLS: ALL CARS must be equipped with a fire proof wall placed between the driver and the motor.

11. IGNITION SWITCHES: All cars must be equipped with a short circuiting device or switch within easy reach of the driver.

12. GASOLINE TANKS: All gasoline tanks must be constructed of at least eighteen (18) gauge material with all joints lapped, riveted and sweated or butt welded.

13. CLUTCH: All cars must be equipped with a hand operated in and out shifting device. A foot operated clutch only is not legal, but may be used with an in and out shifting device.

14. BRAKES: All cars must be equipped with suitable bra device and must pass the inspection on the Contest B 4 wheel brakes allowed.

15. UNDER PANS: All cars must be equipped with an u pan extending from in front of the rear axle tube to a p forward of any foot operated control and must extend side member to side member of the frame.

16. HOODS: All cars must be equipped with a hood w shall consist of a top and two side panels and mus securely fastened to the body or the frame of the car.

17. SAFETY BELTS: All cars must be equipped with a s belt in good working order and drivers will be requ to use them at all times.

18. INSPECTION: All cars must be checked at the trac the Technical Committee when appearing at a track the first time or after being wrecked.

General Rules . . . .

RULE CHANGES: No rule will be changed at the t Notice to every car owner and driver must be mailec by the office.

REFUSAL OF ENTRY: Any entry, even though the petitors are licensed and the car falls within the condi outlined for the meeting may be refused, upon the ex permission of the UNITED RACING ASSOCIATI for reasons such as deportment, financial difficulties actions of the driver, car owners or crew to the best int of the sport, but not otherwise. Any unlicensed ca driver may be refused entry or membership without sta the reason for such refusal.

PHYSICAL EXAMINATIONS: Every driver pass a satisfactory physical examination and be appro by the medical department before he will be perm to drive any car on any track sanctioned by this ciation. The contest board may require any other m ber or official to pass a physical examination.

INTOXICANTS: The use of intoxicants by drivers, owners, mechanics or pit attendants or the regular t officials is strictly prohibited under penalty of immec exclusion by the Contest Board and/or the track doc examination and later fined, suspended or disqualified f the UNITED RACING ASSOCIATION. Any driver has been drinking on the day or night of a race, wil disqualified for thirty (30) days and fined fifty do ($50.00). Second offense, disqualification for the bal of the season. Drinking in the pits by any member or manager will cause that member to be taken from the and the car excluded from further competition that meeting. A policeman in the employ of the promoter be called to assist the Contest Board or Pit Manage enforce this rule.

REGISTRATION:

A. All cars and drivers and officials competing or officia at races sanctioned and conducted under the rules regulations of the UNITED RACING ASSOCIATIO must be registered at the Association office.

B. The transfer of car ownership must be registered at Association office and a Notarized bill of sale r accompany the transfer and registration with the fe required attached thereto.

UNSANCTIONED RACES: No car or driver will be permitted to compete in any unsanctioned race meet without written permission of the Contest Board of the UNITED RACING ASSOCIATION. Any car owner or driver doing so will be subject to investigation and if found guilty will be fined or suspended or both.

REGISTRATION FEES: Car owner, $5.00. Car owner-driver, $7.50. Driver, $5.00. Mechanic, $5.00. Each additional car, $5.00. Officials, $5.00. Photographers, $5.00. Membership fees shall be good for the current year. A member will be considered in good standing until March 5th of the following year.

PLATES, CARDS AND PINS: The car owner will receive his registration plate and a membership card. The plate must be attached to the instrument panel for display at all race meets. A driver or other membership classification will receive the membership card only. Membership pins will be supplied at cost to registered members.

AGE REQUIREMENTS: NO DRIVER or MECHANIC under the age of twenty-one (21) will be eligible for a license. NO PIT ATTENDANT or ANY OTHER PERSON under the age of twenty-one (21) will be allowed in or on the track or the pits during any sanctioned race meet.

NUMBERS: All cars must carry their correct numbers as issued by the office of the United Racing Association. They must be painted in contrasting colors on each side of the oil. Cars failing to comply with this rule will not be allowed to compete. Numbers must be painted not less than 10 inches high and 2 inches wide.

APPEARANCE: All cars must be painted and appear at the track in a clean and presentable condition subject to the approval of the Contest Board, before being allowed to compete. If damaged at the track, a car must be repainted within seven (7) days after returning to competition.

WRECKED CARS: After a wrecked car has been repaired it will not be permitted in competition until the Technical Committee has inspected and OK'd it for racing.

DRIVERS: All drivers must appear in clean white or other suitable uniforms, with helmet to match. Car owners and their pit crews must be dressed in white pants or suitable uniforms. Car owners will be held responsible for the fines of pit crews or drivers not complying with this rule. First offense, $2.00 fine; second offense, $5.00 fine; third offense, $10.00 fine and car barred from the event.

PIT CREWS: The pit crew of each car will be limited to the driver and four pit attendants who will have to remain in the immediate vicinity of their pit during race meets. One pit attendant must be in or with the car at all times while in the pits. Each pit attendant is limited to One Hundred ($100.00) Dollars hospitalization or Doctor's care if not a registered member of U.R.A. Pit pass fee shall be one ($1.00) dollar each and this shall be placed in the Benevolent Fund. There will be no refund of pit pass fees collected.

PIT PASSES: Each car owner, driver and pit attendant will be furnished with proper identification tags, providing they are wearing white pants or suitable uniforms. Pit passes must be worn in a conspicuous manner at all times during race meets. Any one found giving or loaning his pit pass to allow some one in the pits who is not entitled to be there, will cause himself and his car to be barred from further competition on that day. PIT PASSES MAY BE REFUSED CAR OWNERS OR MECHANICS IF THEIR CAR IS NOT PRESENT.

POINT SCORING: Main event only receives points towards championship payoff. Only the first five championship positions will be guaranteed, next five places only by approval of the Board of Directors. Official points will be kept on all races run under U.R.A. sanction between March 1st and December 1st of each year. No points will be allowed on races run between the Months of December and March.

POINT SCHEDULES:

MAIN EVENT ONLY.

1st	1.00
2nd	.8
3rd	.6
4th	.4
5th	.2
6th	.1

One tenth point scored each car from six back to last position. MAXIMUM POINTS that will be scored any car or driver regardless of laps will be 1st, 100; 2nd, 80; 3rd, 60; 4th, 40; 5th, 20. All positions back from 6th, 10 points DRIVERS MUST QUALIFY the car he drives in any main event to score the points won by the car. A car must run one complete lap in any main event to earn a point.

NUMBERS: Numbers will be allotted to cars as they finish in point standings up to fiftieth (50) position. These numbers will be assigned as follows: First ten positions won can not be used by any other car, nor can they be transferred. Numbers from eleven to twenty inclusive may be exchanged with numbers from twenty-one to fifty upon written notice to the office, acknowledged by the Board of Directors in writing. (Car numbers are important. They are the only record the office has for payment of your winnings. The office will not be held responsible for pay checks made out and mailed to wrong owner, when the number scored is on the car illegally.)

Racing Rules . . . .

STARTS: All races will be rolling starts. The pole car has the dress, he may bring the field down fast or slow as he may elect. If the outside front row car willfully jumps the start he does so at the risk of being pulled in and started in the last row. No more than THREE false starts will be allowed. After third false start the offender will be pulled and placed in the last position of the event. The starter's decision will determine the offender.

FOUL DRIVING: Foul driving, cutting the marker for no real reason or unnecessarily bumping, car will lose one position for the first offense, two positions for the second offense, will be paid last position for the third offense. The driver will be fined $200.00 for the fourth offense and barred one year for the fifth offense. The referee's decisions will govern all cases and there shall be no appeal.

PASSING: The starter will give the passing flag to a driver only when he is being overtaken by the lead cars which hold a position in the race better than his. Or on demand of the Contest Board which in its opinion a car is willfully holding back a faster car. Failure to move over will be cause to be pulled from the event immediately.

REPAIRS: Should a car pull out of the field to make adjustments before the start of an event he will have to be back in his position before five (5) laps have been run after the end of the parade lap. At this time the race will be started, regardless of how many cars stop. Only five extra parade laps will be allowed in any event.

CRASH HELMETS: All drivers must wear a crash helmet which meets the approval of the contest board.

GOGGLES: No metal frames on goggles will be allowed.

ALTERNATE CARS: A car dropping out of the heat race for repairs will be replaced by an alternate car.

OVERSIZE MOTORS: Should a motor be measured after running in a race and is found to be oversize, the car will be scored in last position and the driver only will be paid the money won for that position. Under no circumstances will the car or its owner be entitled to any part of the money won. The car and owner will be barred from competition for a period of not less than thirty days. A greater suspension or fine or both may be set by the Board of Directors.

QUALIFYING:

No fast laps previous to qualifying will be allowed on dirt tracks. A $5.00 fine per lap or part thereof will be assessed against the car and deducted from subsequent winnings.

DIRT TRACKS: Take one lap and save two. For your first time out you will be on the timer the second time around or passing over the starting line. For your 2nd and 3rd laps you will be on the timer the first time around or passing over the starting line. FAILURE TO QUALIFY IN YOUR CORRECT ORDER FOR THE FIRST LAP CAUSES YOU TO LOSE THAT LAP AND YOUR FIRST ROUND QUALIFYING POSITION. THE SAME RULE APPLIES TO THE SECOND ROUND OF QUALIFYING.

ASPHALT TRACKS: Take all three laps consecutively. FAILURE TO COMPLETE THREE LAPS IN A ROW, YOU LOSE ONE LAP.

NOTE: Failure to complete a lap after taking the flag on the first or second lap will cause loss of that lap. Within the time limit set for qualifying, all cars ready to run will be given at least one lap on the clock.

QUALIFYING ORDER: Shall be the way you finished the Main and Semi-main events at the previous race at that track. Remaining cars on hand will be qualified in the order they passed through the pit gate on that day. In the event of timing failure or bad ambulance service, etc., all cars will be given the laps available to them. OPENING A NEW TRACK OR THE FIRST RACE OF THE SEASON, THE QUALIFYING ORDER SHALL BE ACCORDING TO CAR NUMBERS ONLY!

HAZARDS: No car or cars will be started or permitted to continue in competition, which, in the opinion of the Referee or Steward or Starter, is a hazard to other cars in the same event.

SPEED: Cars running slowly due to mechanical difficulties will be flagged into the pits after 5 laps.

RESTARTS: There shall be NO RESTARTS unless the track is dangerously blocked and this shall be at the Starter's decision. If the race is stopped, the cars shall be restarted in approximately their GAINED POSITIONS. The lead car may set the pace he desires.

FINISH: When the first five cars have completed the required number of laps in any major event, or the checkered flag has been given, the remaining cars will place according to their positions on the score sheet DURING the lap in which the race was so ended.

CATCH TANKS: On asphalt tracks all cars must carry a standard water catch tank that has been approved by the Association. To protest a car throwing water the protesting driver must pull in with enough evidence to prove his charge. Protesting car will be allowed back in his original position regardless of laps lost while protesting. The offending car will be flagged off the track and paid according to scored position in the official finish. Fine and suspension or both may be set by the Contest Board.

SPECIAL RULINGS: Track management may request that certain types of tires or treads be barred at their tracks or that muffled exhausts will be required due to public hazards and public nuisance ordinances.

MUFFLERS: A standard muffler that has been approved by the Association will be the only type of muffler allowed at tracks where such equipment is required.

LOSS OF MUFFLER: Any car with a disconnected muffler due to broken brackets, separation at the header pipe or any other reason, must pull in immediately regardless of his position in the event. Failure to do so will forfeit any money won in that event.

ALTERNATES: All Main and Semi-main events shall have an alternate car in the lineup. This car shall trail the field and come into the pits after the second lap if no car drops out. Eligibility shall be determined by qualifying time.

INEXPERIENCED DRIVERS: No driver will be allowed in a Trophy Dash or Main event who has never driven a midget auto race car in competition. He must compete in two Semi-main events, heats not considered. He will be started in last position until in the opinion of the Contest Board he is qualified to handle the car.

MECHANICS: Each car entered in a race shall have one registered mechanic with the car at all events.

PROTEST:

All disputes, regardless of their qualifications, will not be recognized unless they are in writing and signed, or on a form designed for this purpose and signed.

A written protest must be accompanied by $25.00. If the protest is valid, the money will be returned. If the protest is void the money will be forfeited and placed in the General Fund. In the case of a mechanical protest, the money will go to the car owner protested if the protest is not valid. If the protest is valid the money will be returned to the party entering the protest.

Contest Board Competitive Committee Approved Regulations

The following rules and regulations have been approved by the contest board of the Central States Racing Association, Inc., and govern the conduct of this meeting.

Please Be Advised That They Will Be Enforced to the Letter.

veryone taking part in any sanctioned Central States Racing Association, Inc., race or contest is expected to become familiar with e official competition rules. Ignorance of the rules will not and cannot be accepted as a plausible or reasonable excuse for any and all violations or infractions.

1. **ENTRY.** Shall consist of a combination of car and driver. Entry ank must be completely filled out, including the duplicate and complete a given on every question. ENTRIES POSITIVELY CLOSE at Midnight days prior to the race meet described on this entry blank or stated to contrary on the title page. Entries may be made by telegraph. Should eligible combination car and driver arrive at the track without proper ry, commonly called a post-entry, they may be eligible to compete in the 's activities, after payment of $2.00 cash, post-entry fee, but will only entitled to derive 80% of the advertised purse of this race. The balance % of said combination winnings placd into the CSRA Contingency Fund, ar securing permission of all eligible entrants.

2 ELIGIBILITY—All drivers and cars must be registered by the Contest ard CSRA. No unregistered driver or car will be allowed to compete in association event unless he has registered as required by this rule, and in good standing and eligible. Entries may be made by non-registered vers and cars. If you are not registered, in filling out this entry blank ite in the space provided for the car and driver registry, NOT REGIS-RED. A licensed mechanic of this organization MUST be in attendanc h each car.

3. FEES FOR REGISTRATION—Fee for registration up to May 29th any one year are on the following annual license based schedule: Drivers nual license fee $7.50 Owners annual registration fee $7.50. Mechan-an's annual registration fee $5.00 After May 29th the fee for registra-n can be set at a sum consistent with the past known record of the ap-cant, but in no case can it be less than the fees set for members on the nual schedule. Cash bonds evidence of good faith covering the fulfilling all requirements for CSRA membership may be required of some. These h bonds in no case will be less than $200.

4. CANCELING EVENTS—The management reserves the right to cancel event unless a satisfactory number of entries have been received by date on which entries close. In the event of cancellation, notice will be de to all entries by wire.

5. CAR NUMBERS—The assignment of car numbers is made by the con-t board office. The representative shall have full authority on the issuing temporary numbers, for those who have not received their official number m the contest board office, for that particular race covered in this entry nk. The champion drivers are honored with the low numbers, according the official standing of the previous year. Numbers assigned covering the nporary race issue will be made by the representatives to avoid confliction, that no two cars will display similar numbers. These numbers are to be inted on both sides of the tail, and at least ten inches high at this loca-n. It is suggested, though not mandatory, to augment the numbers on the l, with the official assigned number at a place on the right side of the wl at a distance near the center. The number need not be more than ht inches.

6. STARTING.—Will be from a flying start. The entry list will be qual-d by the OFFICIAL STARTER and then arranged in seniority with the test car first unless stated contrary on title page. Similar time established the official qualifying trials by partcipants will result in those, by reason qualifying first, being honored and granted preference over the other trants who may later equal their qualifying time trials, in making up the rting positions in the day's races. Only the pole car driver has the priv-ge to start or refuse the start as well as stop the event in which he takes rt in. All other competitors must be guided accordingly.

7. NUMBER OF LAPS FOR QUALIFYING—A maximum limit of three s will be allowed any one qualifier for this purpose, if the entrant so sires and it may be necessary in order to become eligible to the day's sub-quent event or events. The fastest lap of the three to be considered as cial by the judge or board representative, with no added attempts granted der any condition later during the day's program. The date and hour of alifying trials will be stated on the face of the entry blank. Drivers alifying after the OFFICIAL HOURS lose positions to which trials would title them. Races under the association WILL NOT BE PERMITTED TO HELD UP FROM THE SCHEDULED TIME OF BEGINNING caused by ays on the part of drivers, or car owners. No other car will be per-tted on the track during the while a contestant is attempting to qualify. ivers must go to the end of starting line in event more than the allotted mber of trials may be required for qualifying. In event a driver secures rmission to qualify more than one car, then the subsequent car other than original car he drives in the initial events must go to the end of starting sitions in the event the car is eligible to.

8. RESTARTED RACES OR EVENTS—Prior to the completion of one mplete lap, should a restart be necessary, cars will start from their orig-al positions. However, should a restart be required after the completion one or more laps, then a restart will be made in accordance to positions prior lap, and the remaining laps to be run constitute the event.

9. STARTER IN CHARGE—The starter will be in complete charge of the nduct of starting the race. He must be an accepted person to the contest ard office or to the board representative in charge as to fitness and ability. e starter's principal duties consist of: To satisfy himself that proper of-ials, scorers, judges are provided. To see that all cars carry proper iden-ication numbers, to bring the cars to the starting line in uniform and stomary order and only then to start them. To carry out instructions from ard representatives in charge.

Shown here are the rules and regulations of the Central States Racing Association--the governing body for dirt track and midget racing in the Midwestern United States.

10. **EXAMINATION, MOTOR LIMITATION**—All cars may be examined by the official representative, his aid or committee as to engine measurements preceding the race. The classes in motor limitations may be in groupings of "A" and "B." Car engines or motors must not be less than 150 c.i.d. for the speedway car division. In the midget division limited to conventional midget racing engines up to 105 c.i.d. overhead valves, and 140 c.i.d. flathead motors. The limitations on motorcycle engines shall be no greater than 30;50, with absolutely no other restrictions or limitations to apply to racing motorcycle motors. A racing sponsor or promoter who is desirous of limitating the engine sizes for his event to any smaller than the maximum motor sizes is permitted to do so provided the CSRA executive offices approves the special limitation which the sponsor chooses to permit in his competitions, two weeks prior to the race meeting, and it is carried on the face of the official entry blank. Converted aviation motored cars are permissible as defined by the racing sponsor and stated on the face of the entry blank, but CSRA has set no restrictive limitations on sizes of this type engine. Class "A" is representative of both championship and non-championship but these events must be approved by the contest board on application, with the conditions under which they are to be run stated clearly on the face of the entry blank. "SUPERCHARGERS" are permitted on engines or motors up to and including 151 c.i.d. only for speedway events, and do not include midget motored cars or motorcycles. Engines with superchargers attached are permitted only to participate in class "A" meets. Wheel base of all eligible speedway cars must be no less than 80 inches nor in excess of 114 inches hub to hub maximum. Quality of painting and workmanship will be taken into consideration in determining the degree of preparation that has been made for the contest. As a general rule cars which are makeshift in appearance are makeshift mechanically and vice versa.

*Underpan extending from a place near motor support and near closest distance or rear axle is demanded.

*Fires dash of appropriate metal is necessary.

11. MECHANICAL REGULATIONS—Each eligible car must have suitable and safe brakes at least on the rear wheels, that are in good working order. Steering wheel spiders must be either steel or bronze. Steering wheel mechanism, knuckles and spindles must be passed upon and approved by the CSRA representative, his aid, or Technical Judge. Exhaust pipes must be extended beyond rear of driver's seat. Metal fire dash is compulsory as is a under pan under the driver's compartment. An appliance must be installed on all registered cars, that should a rear axle break, the appliance will safely hold the wheel on the car. Size and fitness of rear axles will be taken into consideration by the OFFICIAL REPRESENTATIVE as to the matter of safety.

12. PAYMENT OF PRIZE MONEY—Prize money will be paid to entrants of winning cars or drivers immediately after the finish of the race meetings excepting in case of protest. An entrant protesting to the official board representative prize money, will be compelled to accompany the protest with a deposit of $10.00. If the finding of the contest board is in favor of the protestant, the $10.00 deposit accompanying the written petition will be returned to the protestant. If, however, the finding of the board is not in favor of the protestant, the $10.00 plus 20% of enrant's funds held in escrow by reason of his filing written protest for the consideration in question, after reasonable expenses to defray the cost of the board hearing the action on the protest is deducted, shall be added to the CONTINGENCY FUND. Any and all protest must be made immediately after the disputed event ,before prize awards are paid off.

***NOTE—In case of protest against motors exceeding the limits set for competitions, rule No. 12 has been amended whereby the car owner or mechanic in charge of the car whose motor is being protested, receives the amount of protest fee deposited, $10.00 upon tearing down the motor and proving to the contest board representative, his aid or special appointed official, the car's engine conforms with the maximum association limit.**

13. DISQUALIFICATION—No entrant shall have any claims for damages, expense, or otherwise, against management or any of its officials or representatives or CSRA board by reason of disqualification of either car or driver. Contest board action and decisions are final

14. PIT AND GRANDSTAND RULES—Each entrant will be allowed four pit tags and complimentary tickets. Proper deportment in each individual pit is charged to the driver and licensed mechanic. They will see that pitmen are selected who serve him, and that all pitmen are instructed in the following:

1. Pitmen must remain as far as possible in their own pit. Promiscuous visiting will not be permitted.
2. No one will be allowed on track or in the pits without the proper tag, which must be visible at all times.
3. There must be no standing or sitting on any rail. During the course of a race, ONLY ONE attache of each competing car will be allowed in front of HIS OWN RESPECTIVE PIT. All others must remain in BACK OF THE PITS during the actual race.
4. Cars must be kept arranged in an orderly manner parallel to the track, if possible, and close to the pit.
5. Pitmen must keep their eyes on their own car at all times, ready to help their driver. That is what they are there for and nothing else.
6. Neatness and cleanliness are required in pitmen, as well as drivers. Grease-soaked clothing WILL NOT BE PERMITTED. A cash prize will be paid at each race to those whom spectators in the stand by favorable reaction, deem the best appearing combination of car and crew at this particular race.
7. Drinking by anyone in the pits or connected with the pits either before or during the race, of intoxicating liquors will not be tolerated by CSRA officials.

15. HEATS—Where more than the maximum number of starters are entered, and especially where the day's program consists of a number of sprint events, the management reserves the right to hold heat or elimination trials other than specified under schedule of events, distances and conditions to be approved by the management.

16. **APPEARANCE**—All drivers, car owners and pit crews must be clothed in clean white coveralls or similar appropriate uniforms. Carelessness will not be tolerated. The people in the grandstand are entitled to see clean, well-dressed performers. It is their money that makes professional racing possible. Grease-soaked clothing will not be tolerated.

17. All drivers will be guided by the following signals: Any driver who fails to heed any flag will be promptly ruled off the track. No exceptions.

GREEN FLAG—Start, course is clear.
*YELLOW FLAG—Caution; watch out for conditions ahead; get car under control and hold position.
RED FLAG—Stop, race is halted.
BLACK FLAG—Stop next lap for consultation to driver designated.
KING BLUE FLAG—Competitor is trying to overtake you.
WHITE FLAG—You are entering your last lap.
CHECKERED FLAG—You are finished.
*This is a warning signal and mandate that all cars be brought under
control immediately.

18. Every driver and riding mechanic must wear an approved type of crash helmet at all times during a race and practice session. Shatter proof goggles MUST be worn as well. Every driver must pass a stated physical examination to become eligible to CSRA sanctioned races, on demand.

19. **LIABILITY**—The entrant in signing this agreement elects to use said track at his or its own risk thereby releases and discharges said organiser, the race sponsor and the Central States Racing Association, Inc., CSRA Contingency Fund, Inc., together with their successors, officers, agents, and employees, from all liability from personal injuries that may be received by said entrant, and from all claims and demands for damages to personal property or employee growing out of or resulting from the race or races or events contemplated under this entry blank or caused by any construction or condition of the track of said organiser.

20. Should any condition arise which is not covered by CSRA or special rules, then official representatives of contest board shall have the immediate right to act in all such matters.

21. Annual license card must be presented on request of authorised test board official.

22. **UNSANCTIONED RACE MEETINGS**—THE ACT OF TAKING PA COMPETING, AIDING, OR ACTING AS AN OFFICIAL BY ANY MEME OF THE CENTRAL STATES RACING ASSOCIATION, ON UNLICEN TRACKS, SPEEDWAYS OR COURSES, SHALL AUTOMATICALLY QUALIFY AND SUSPEND WITHOUT ACTION OR FORMAL NOTIFI TION BY THE CONTEST BOARD, AND SUCH DISQUALIFICATION SUSPENSION SHALL REMAIN IN EFFECT UNTIL REMOVED BY F MAL ACTION OF THE CONTEST BOARD AND BOARD OF DIRECTO

AN ENTRY IN ANY UNSANCTIONED CONTEST, OR AN AUTH IZED ANNOUNCEMENT IN PUBLIC THAT AN ENTRY HAS BE OR WILL BE MADE, SHALL BE DEEMED SUFFICIENT CAUSE IMMEDIATE DISQUALIFICATION BY THE CONTEST BOARD OF OWNER, ENTRANT, DRIVER AND CAR, OR ANY OR ALL OF THEM

23. **FAILURE TO START**—Should a Driver fail to appear or start shall, unless excused by the representative or Promoter, be reported to Contest Board. Penalty may include fine, suspension or disqualification the circumstances warrant.

24. **UNREGISTERED OR DISQUALIFIED DRIVER ON LICEN TRACK**—The owner or lessees of a licensed track who permit an unre tered or disqualified Driver to use said track shall be suspended and t track license cancelled for such time as the Contest Board may determine.

25. Racing sponsors desirous of making regulations in connection their particular race meeting beyond those covered by the rules, shall sul such supplementary regulations in advance of their printing to the CS Executive offices

#26 Rules and regulations are applicable to speedway, midget raci cars and motorcycles. Where reference specifies either speedway or midg cars, drivers or competition, and applies only to motorcycle racing riders, then rules and regulations shall apply to motorcycle as well, wha ever the case may be.

ESSENTIAL RACING CAR CONSTRUCTION DETAILS

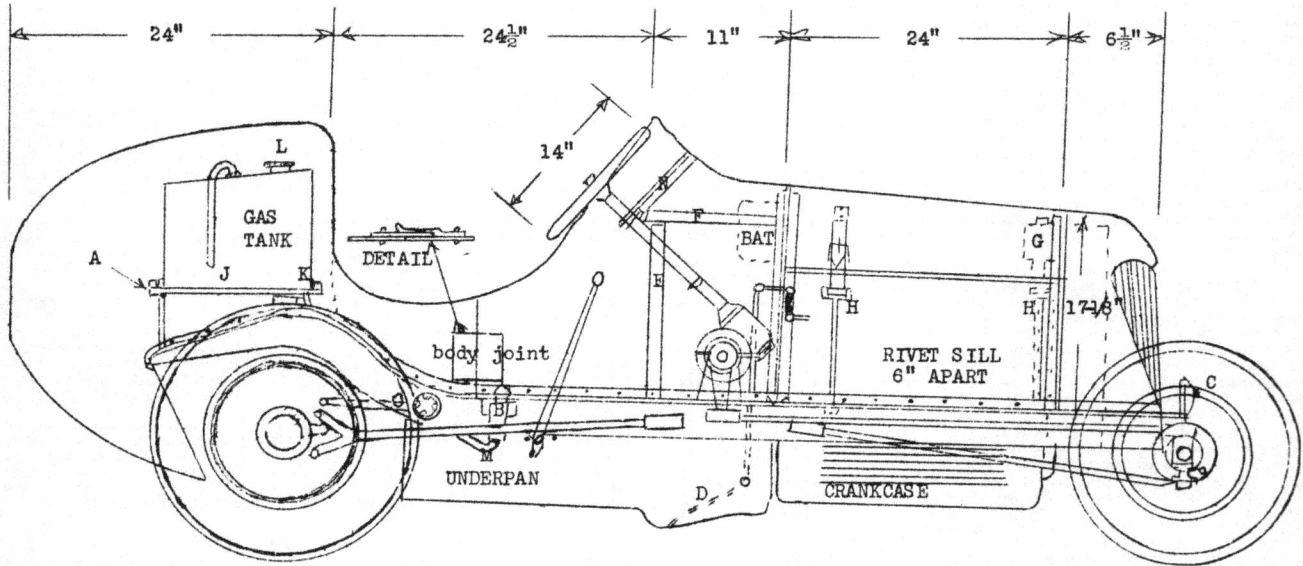

A - Hold-down lugs.
B - Air Pump.
C - Hartford shock absorbers.
D - Gas pedal.
E - 1½ in x 1/8 in. for mid-
    get or heavier stock for
    big car.
F - 1 in. x approx. 11 in x 1/8
    or heavier stock for big car.
G - Radiator tank & core.

H - Hood fasteners - two type.
J - Air line to pressure pump.
K - 1/2 in gas line to carburetor.
L - Air tight gas tank cap.
M - Fasten tee to connect with air
    gauge on instrument board.
N - 1 x 1 x 1/8 in. Angle iron, or
    use heavier stock for big car.

NOTE: DO NOT SCALE THIS DRAWING.

Above drawing for two-spring 'round nose' frame.

The drawing shown is for a Midget Racing Car. Proportions for a Dirt Track Car are similar but larger dimensions are used throughout.

When one undertakes the construction of a racing car, whether it be a 'three-quarter' (dirt track, full-size car) or a midget racing car, the first consideration is the power plant. In dirt track cars many combinations are used, full racing engines, and semi-stock or converted engines, with the count about equally divided between the two classes. Among the midget cars the owners and builders are getting away from conversions of stock engines (with a few notable exceptions), and are specializing more and more on strict racing engines. Dirt track cars generally use Miller, Offenhauser, or Meyer-Drake engines (all quite similar) in the custom-built racing engine class, and Ford V8, Mercury. Model B Ford, and a very few Hissos, Cragars, Rileys, and similar old engines in the semi-stock engine class. The Fords, Mercuries, etc., are all sufficiently rebuilt and equipped with supplementary equipment that they are closer to strict racing engines than to stock engines.

Midget racing cars generally use one of the following engines:

1. MEYER-DRAKE (Offenhauser) midget engine (custom racing engine)

2. FORD V8 '60' (stock 60 h.p. Ford V8 engine, not in production, for which a great deal of high speed supplementary equipment is available from various manufacturers)

3. FORD-FERGUSON (tractor) engine (Industrial & tractor engine of 120 c.i.d., still in production. OHV available from Roof, Dreyer, and others)

4. DRAKE (special conversion of 2-cylinder motorcycle engines)

5. J. A. P. (Twin-cylinder English motorcycle engine of 61 c.i.d. which gives 80 B H P with alcohol fuel)

6. WILLYS 4 (jeep engine) not widely used. (O.&.V. set-ups are supplied by Roof and Pacific Metals)

In the early days of midget racing many different passenger car engines were rebuilt for use in midgets, but this practice has ceased almost entirely. In those days designers who liked to experiment did such things as cut a Marmon eight in half, cut a Pontiac six in half, cut a Model 'A' Ford engine in half. Each of these engines were made to run, some of them efficiently, but the grief involved was hardly worth the trouble. Other engines used in the early days of midget racing were the Henderson 4 (motorcycle) engine, and the Saxon (ancient small automobile engine).

Midget racing has become so highly specialized, and has advanced to such a degree, that converting and rebuilding any out-of-the-ordinary engines is not recommended, and should not be considered. Any one of the six engines listed above, however, will give high performance and creditable service.

From the standpoint of 'availability', and expense, the Ford V8 '60' is one of the best engines available for use in midget racing cars. The alterations which are necessary are few, and the engine is highly efficient and will give good performance. It is an ideal engine for the novice car builder to choose because, without the expenditure of much money, one can make improvements in the engine which will increase its efficiency. A ground camshaft can be installed, the compression increased, (by the use of higher compression heads) and a twin-carburetor manifold installed. Some owners have reversed the porting system, (intake to exhaust, and exhaust to intake) in an effort to make the engine run cooler. A great deal of supplementary high speed equipment is now available from various manufacturers for the V8 '60', and some combinations of this equipment give highly creditable performances. Some manufacturers claim as high as 118

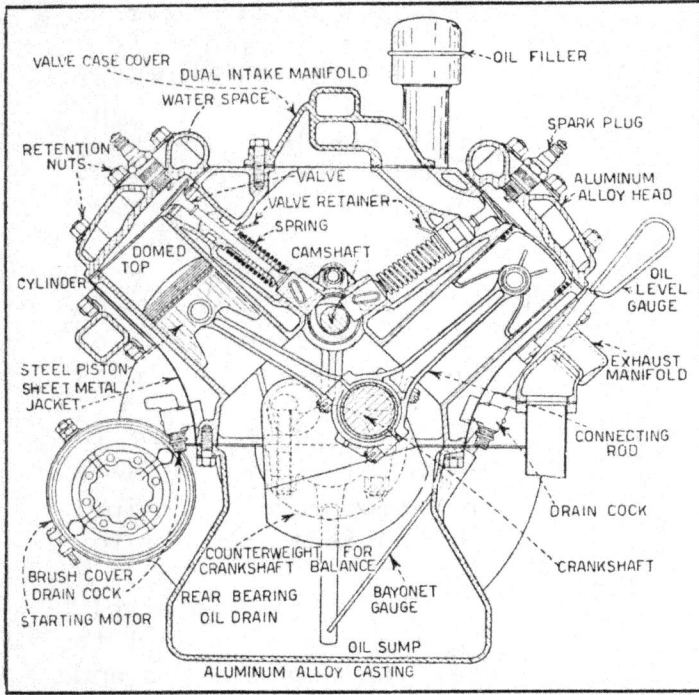

TRANSVERSE SECTIONAL VIEW.

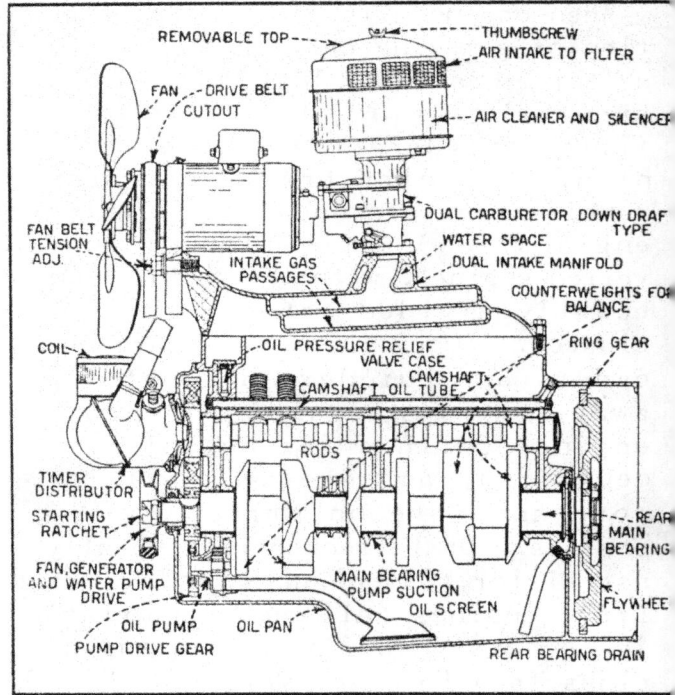

Labels (left diagram):
VALVE CASE COVER
DUAL INTAKE MANIFOLD
WATER SPACE
OIL FILLER
RETENTION NUTS
VALVE
VALVE RETAINER
SPRING
CAMSHAFT
SPARK PLUG
ALUMINUM ALLOY HEAD
DOMED TOP
CYLINDER
OIL LEVEL GAUGE
STEEL PISTON SHEET METAL JACKET
EXHAUST MANIFOLD
CONNECTING ROD
DRAIN COCK
BRUSH COVER DRAIN COCK
STARTING MOTOR
COUNTERWEIGHT FOR CRANKSHAFT BALANCE
REAR BEARING OIL DRAIN
BAYONET GAUGE
CRANKSHAFT
OIL SUMP
ALUMINUM ALLOY CASTING

LONGITUDINAL SECTIONAL VIEW.

Labels (right diagram):
REMOVABLE TOP
THUMBSCREW
AIR INTAKE TO FILTER
FAN
DRIVE BELT CUTOUT
AIR CLEANER AND SILENCER
FAN BELT TENSION ADJ.
DUAL CARBURETOR DOWN DRAFT TYPE
WATER SPACE
DUAL INTAKE MANIFOLD
INTAKE GAS PASSAGES
COUNTERWEIGHTS FOR BALANCE
COIL
OIL PRESSURE RELIEF VALVE CASE
CAMSHAFT
RING GEAR
CAMSHAFT OIL TUBE
RODS
TIMER DISTRIBUTOR
STARTING RATCHET
FAN, GENERATOR AND WATER PUMP DRIVE
MAIN BEARING
REAR MAIN BEARING
OIL PUMP
PUMP DRIVE GEAR
OIL PAN
PUMP SUCTION OIL SCREEN
FLYWHEEL
REAR BEARING DRAIN

FIG. 1.   STOCK FORD V8 - 60 - H.P. ENGINE.

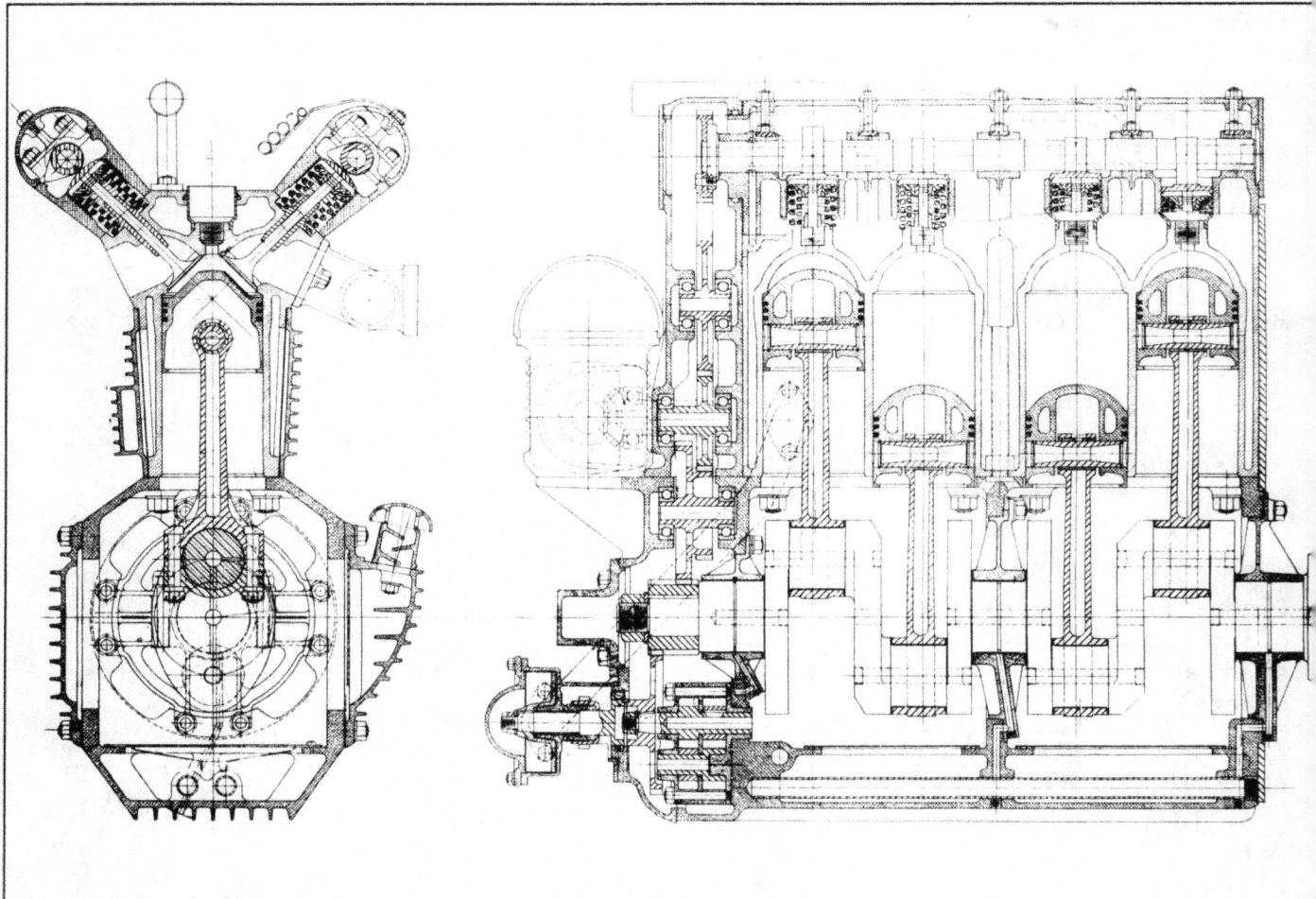

FIG. NO. 2.   MEYER-DRAKE MIDGET RACING CAR ENGINE (COMMONLY REFERRED TO AS THE OFFENHAUSER).   TRANSVERSE SECTION (LEFT) CLEARLY SHOWS THE DOUBLE OVERHEAD CAMSHAFTS AND VALVE SYSTEM.   NOTE DOMED PISTONS AND SHAPE OF COMBUSTION CHAMBER.

B H P for their combinations when used on a V8 '60'. The Offenhauser midget engine gives about 125 B H P. (See Figs. 1 & 2)

If the builder is considering building a dirt track car (three-quarter car), or a track roadster (sometimes called 'hot rods') the V-type eight cylinder engine is by far the most popular and, at present, the most widely used. As with the midget V8s a great deal of extra equipment of a high-performance nature is available. Most of this equipment is built to high standards and will give good service. Mercury engines are the most widely used, and many records have been made with this type of power plant. (See Fig. 3)

The 'flat head' in-line engine has always been a temptation for car owners to build, but it has not always proved to be the most satisfactory, for it is a motor which is very difficult to keep properly cooled. The valves are located so close together on a 'flat head' that there is insufficient room for the cooling water to circulate around them. Some of the drivers have been placing solidified carbon dioxide ('dry ice') in front of the radiator to keep it cool in a long race. There is also the problem of 'hopping up' this type engine to compete with some of the faster cars. This is a rather costly job, and the fear of throwing a rod and ruining the engine is ever-present. (See Figs. 4 & 5)

A few of the problems in building up a good 'flat head' race car follow: one has to obtain an engine on which the cylinder head has the proper compression ratio, and the shape and design of the combustion chambers gives the utmost efficiency. The valve ports of the cylinder block will have to be polished, a new camshaft, ground to racing standards, installed, new oil pump and water pump added, in order to give proper oil pressure and proper water circulation.

Another problem is weight. The entire car should be made as light as is commensurate with safety, as dead weight obviously taxes the efficiency of the engine.

Upon selecting an engine and deciding upon its use, and subsequently stripping it down preparatory to rebuilding, it would be advisable to take the block to one of the larger automotive shops where extensive re-building is done and have the stripped block boiled in the caustic solution which is used before rebuilding all engines. This cleans the block thoroughly, getting rust, corrosion, metal particles, grit, grime, etc., out permanently. After this cleaning process the block can be carefully inspected prior to reboring, honing, porting, and any necessary cutting that has to be done. (See Fig. 6)

SUPERCHARGERS

Superchargers (sometimes called 'blowers') are used in an effort to give a greater capacity of fuel mixture to the combustion chambers, thus resulting in a larger, more powerful, explosion. They are also supposed to atomize the fuel vapors to a higher degree, resulting in a more volatile mixture. The pressurized mixture thus forced into the engine is supposed to distribute itself more evenly. Theoretically these factors should combine to provide maximum power and smooth operation of the engine. Superchargers have been used in England and on the continent of Europe extensively, and over there their efficiency and operation has been developed to a very high degree. Engines of very small cubic inch displacement have been engineered to turn to very high r.p.m. thus creating a much larger amount of horse-power than conventional motors of twice (and even more) the displacement. This development was brought about, particularly in Great Britain, because the taxation of motor vehicles has been based on

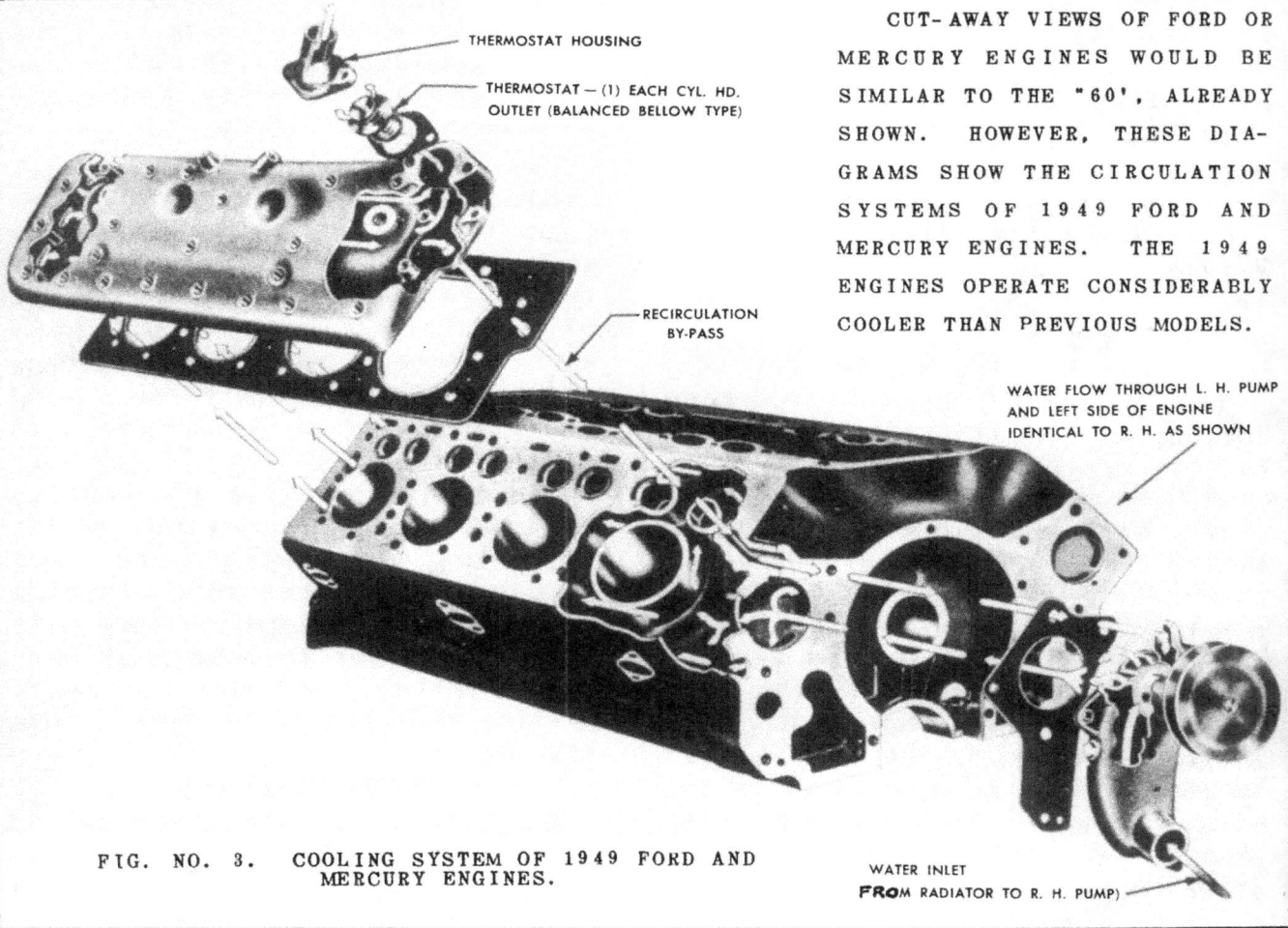

THERMOSTAT HOUSING

THERMOSTAT — (1) EACH CYL. HD. OUTLET (BALANCED BELLOW TYPE)

RECIRCULATION BY-PASS

CUT-AWAY VIEWS OF FORD OR MERCURY ENGINES WOULD BE SIMILAR TO THE "60", ALREADY SHOWN. HOWEVER, THESE DIAGRAMS SHOW THE CIRCULATION SYSTEMS OF 1949 FORD AND MERCURY ENGINES. THE 1949 ENGINES OPERATE CONSIDERABLY COOLER THAN PREVIOUS MODELS.

WATER FLOW THROUGH L. H. PUMP AND LEFT SIDE OF ENGINE IDENTICAL TO R. H. AS SHOWN

WATER INLET FROM RADIATOR TO R. H. PUMP)

FIG. NO. 3.   COOLING SYSTEM OF 1949 FORD AND MERCURY ENGINES.

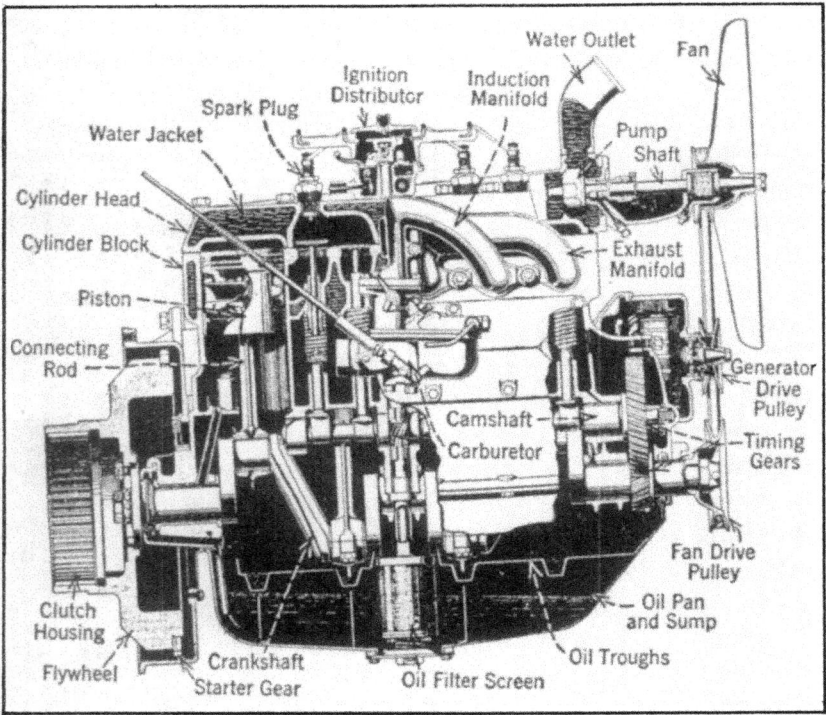

Water Outlet

Fan

Ignition Distributor

Induction Manifold

Spark Plug

Pump Shaft

Water Jacket

Cylinder Head

Cylinder Block

Piston

Connecting Rod

Exhaust Manifold

Generator Drive Pulley

Camshaft

Carburetor

Timing Gears

Fan Drive Pulley

Oil Pan and Sump

Clutch Housing

Flywheel

Crankshaft Starter Gear

Oil Filter Screen

Oil Troughs

LONGITUDINAL SECTIONAL VIEW

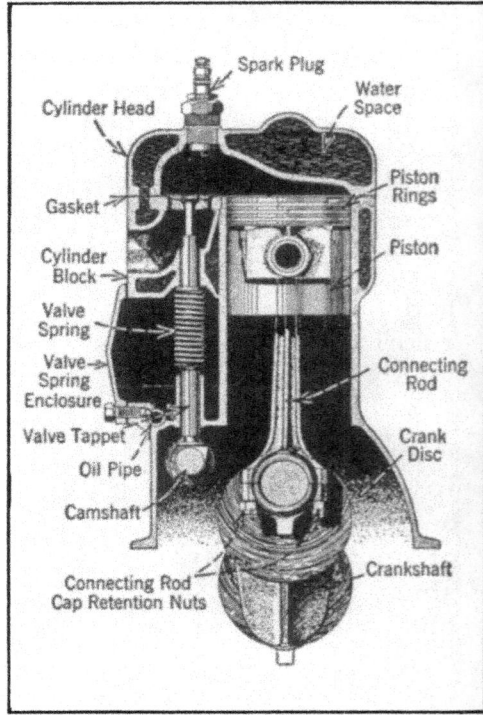

Spark Plug

Cylinder Head

Water Space

Gasket

Piston Rings

Cylinder Block

Piston

Valve Spring

Valve Spring Enclosure

Connecting Rod

Valve Tappet

Oil Pipe

Crank Disc

Camshaft

Connecting Rod Cap Retention Nuts

Crankshaft

TRANSVERSE SECTIONAL VIEW

FIG. NO. 4.   STOCK FORD MODEL A ENGINE.

MODEL "A" ENGINES WERE WIDELY USED IN BOTH TRACK JOBS AND "SOUPED UP" ROADSTERS PRIOR TO THE WIDESPREAD POPULARITY OF THE V8.   SUPPLEMENTARY EQUIPMENT SUCH AS OVERHEADS, CAMS, ETC., WAS AVAILABLE PRIOR TO THE WAR.

14

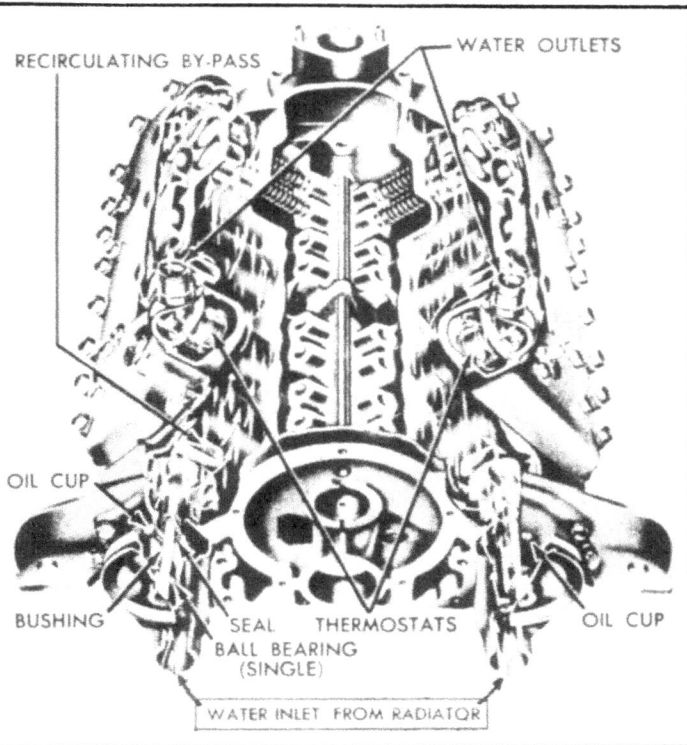

FIG. 4A. ANOTHER VIEW OF THE WATER CIRCULATING SYSTEM OF 1949 FORD AND MERCURY V8 ENGINES.

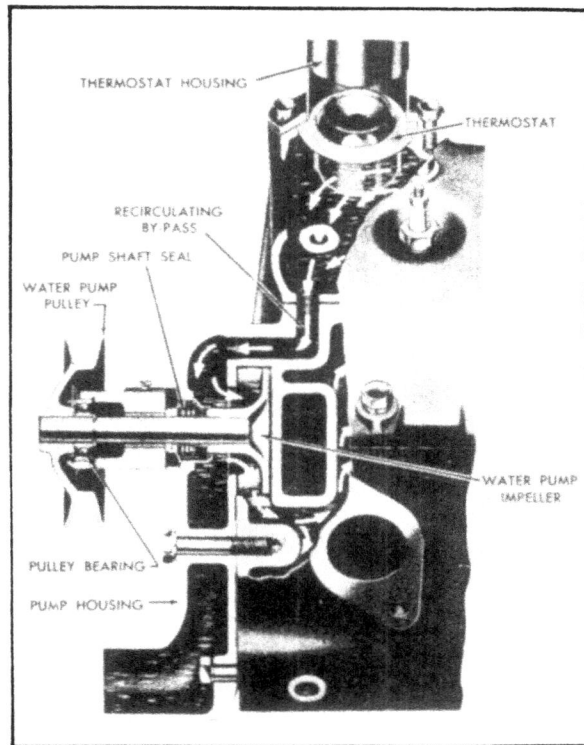

FIG. NO. 4B. DETAIL OF IMPROVED WATER PUMP ON THE 1949 V8s.

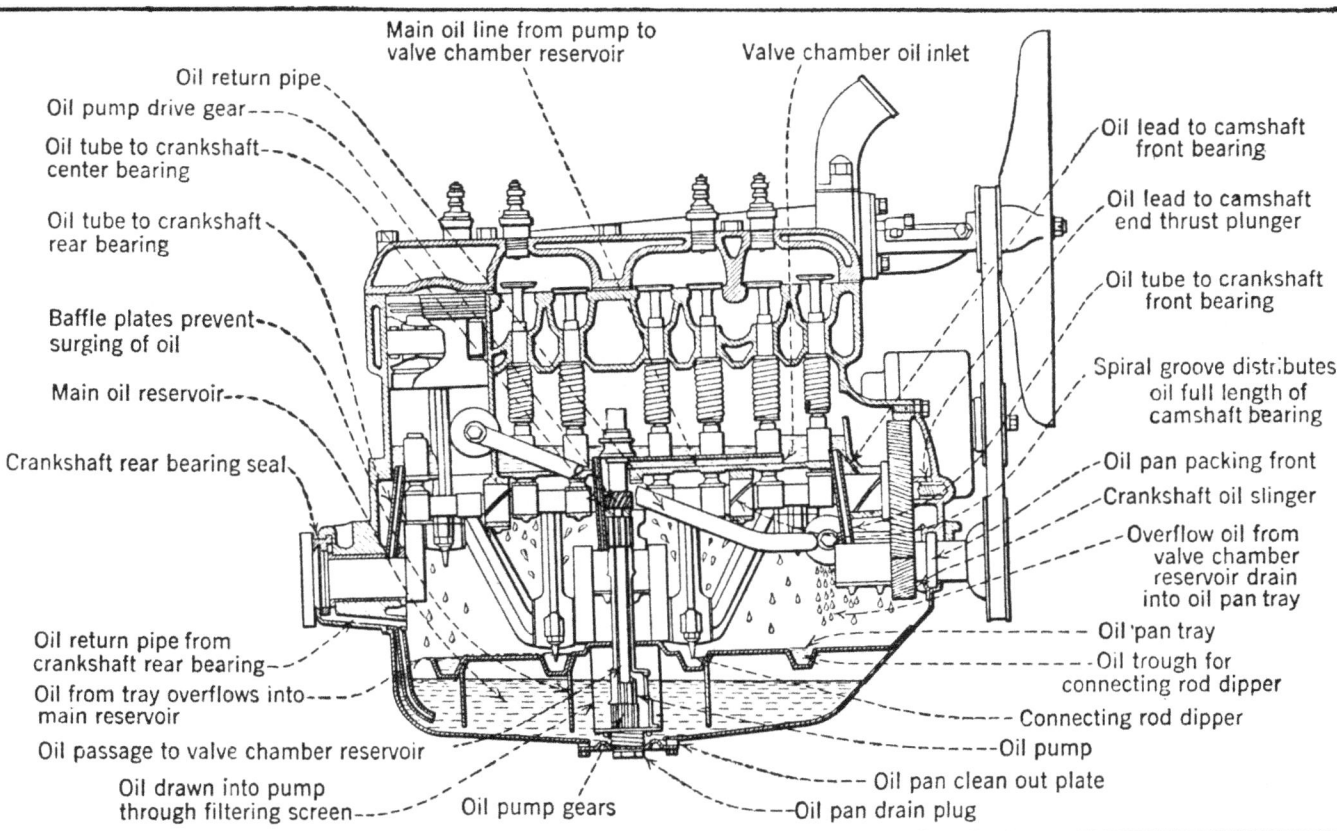

FIG. NO. 5. SECTIONAL VIEW OF FORD MODEL A ENGINE SHOWING METHOD OF LUBRICATING INTERNAL MECHANISM BY COMBINED PUMP AND SPLASH SYSTEM.

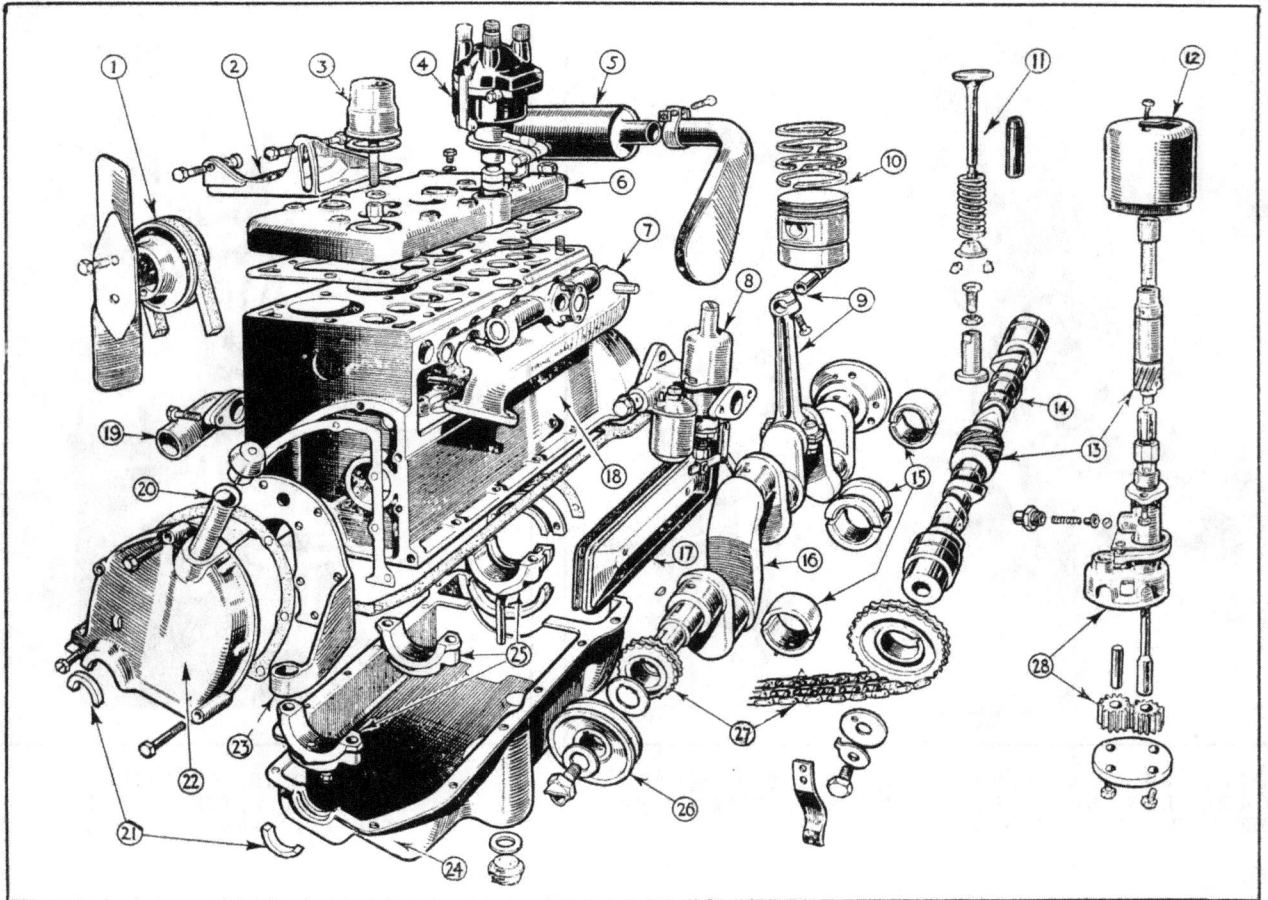

FIG. NO. 6. THIS "EXPLODED" VIEW OF A SMALL 4-CYLINDER FLAT-HEAD ENGINE ACTUALLY SHOWS A BRITISH POWER PLANT. HOWEVER, THE FUNDAMENTAL PRINCIPLES OF ALL INTERNAL COMBUSTION ENGINES ARE THE SAME AND THIS IS A PARTICULARLY CLEAR EXAMPLE OF AN ENGINE'S ESSENTIAL COMPONENTS. SHOWN ARE: (1) FAN & PULLEY; (2) GENERATOR MOUNTING BRACKET; (3) WATER OUTLET FROM CYLINDER HEAD; (4) DISTRIBUTOR; (5) AIR CLEANER; (6) CYLINDER HEAD; (7) INTAKE & EXHAUST MANIFOLDS; (8) CARBURETOR; (9) CONNECTING ROD & WRIST PIN; (10) PISTON & RINGS; (11) VALVE ASSEMBLY; (12) OIL PUMP FILTER & COVER; (13) SPIRAL GEARS FOR OIL PUMP & DISTRIBUTOR DRIVE; (14) CAMSHAFT; (15) MAIN BEARING INSERTS; (16) CRANKSHAFT; (17) VALVE COVER; (18) CYLINDER BLOCK; (19) WATER OUTLET FROM BLOCK; (20) OIL FILLER (TO CRANKCASE); (21) FRONT CRANKSHAFT PACKING GLAND; (22) TIMING GEAR COVER; (23) FRONT ENGINE MOUNT; (24) CRANKCASE; (25) MAIN BEARING CAPS; (26) CRANKSHAFT FAN PULLEY; (27) CAMSHAFT DRIVE; (28) OIL PUMP.

FIG. NO. 7. ROOTS BLOWER USED ON 8CTF MASERATI INDIANAPOLIS RACING CAR. THE ROTORS ARE STEEL.

16

the bore of the cylinder (total piston head area) of the engine. Thus, of necessity, the British made very small engines do the work of large ones, and 'get around' just as fast in doing it.

Superchargers can be divided into two basic classes: the Positive Displacement type, and the Centrifugal type. In the former class are two different designs, the Roots, and the Zoller. (See Figs. 7, 8 & 9)

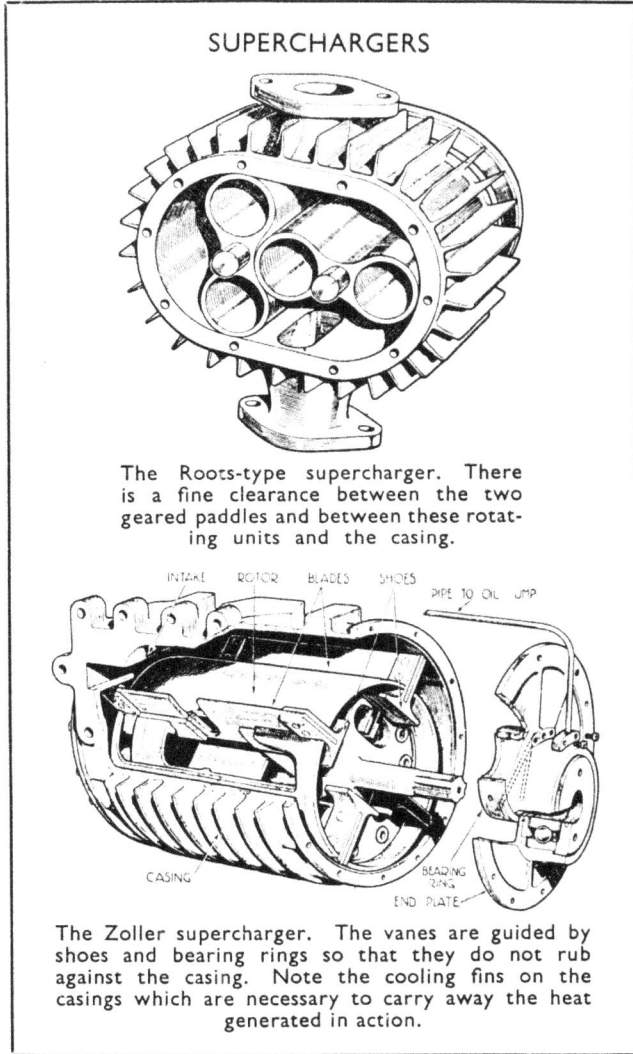

SUPERCHARGERS

The Roots-type supercharger. There is a fine clearance between the two geared paddles and between these rotating units and the casing.

The Zoller supercharger. The vanes are guided by shoes and bearing rings so that they do not rub against the casing. Note the cooling fins on the casings which are necessary to carry away the heat generated in action.

FIG. NO. 8. TWO TYPES OF POSITIVE DISPLACEMENT SUPERCHARGERS: ROOTS (ABOVE) AND ZOLLER (BELOW).

The Roots type blower contains two rotors turning within a case. These rotors turn in opposite directions to each other. They are shaped similar to the figure '8' so that as they turn the wider part of one fits into the narrower part of the other. They are geared outside the case by a pair of closely meshed gears. The clearance between the rotors and the inner surface of the case, and between the rotors themselves is of the closest tolerance. They take air in through a relative-

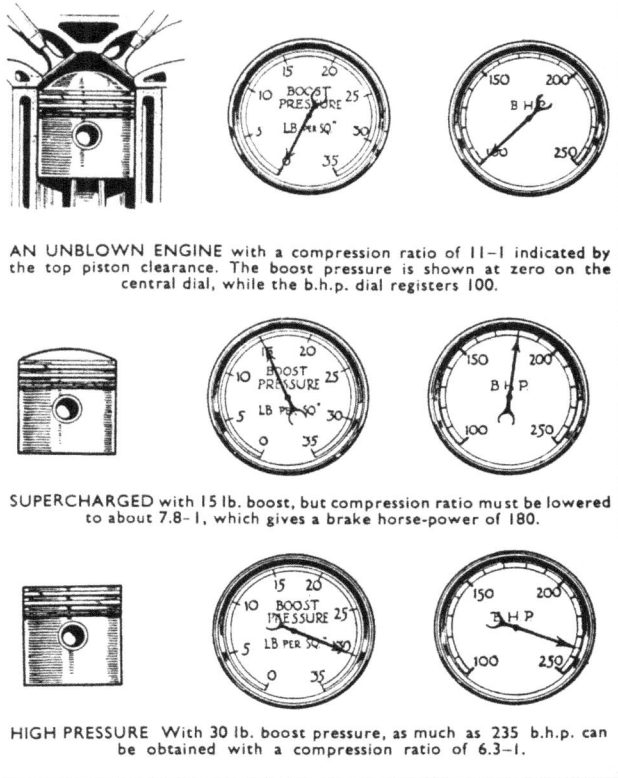

The Relation of Compression Ratio, Supercharging Boost and Engine Horse-power

AN UNBLOWN ENGINE with a compression ratio of 11-1 indicated by the top piston clearance. The boost pressure is shown at zero on the central dial, while the b.h.p. dial registers 100.

SUPERCHARGED with 15 lb. boost, but compression ratio must be lowered to about 7.8-1, which gives a brake horse-power of 180.

HIGH PRESSURE With 30 lb. boost pressure, as much as 235 b.h.p. can be obtained with a compression ratio of 6.3-1.

FIG. NO. 9. NOTE THAT COMPRESSION RATIO IS ACTUALLY DECREASED AS PRESSURE IS INCREASED.

ly large port on one side of the case and force it out through a smaller port on the opposite side of the case. The exhaust port being smaller causes the air to be pushed out with considerable force. As the speed of the rotors is increased this 'push' is multiplied. The blower is geared directly to the engine, and so its speed is in direct proportion to that of the engine. The first successful supercharging of racing cars was with a Roots type blower which has the advantage of no heavily loaded bearings, needs no great amount of lubrication, and is quite reliable.

Positive displacement blowers of the Roots type are efficient at low speeds and give an engine excellent low speed acceleration. Efficiency at higher speeds falls off rapidly - both factors directly opposite to the characteristics of the centrifugal type. Hence the Roots type blower must be large, and can seldom be run at more than engine speed. Speeds much over 5,000 r.p.m. are likely to lead to trouble unless such facilities (and financial reserve) are available, such as were available to the builders of the German Grand Prix cars.

The Zoller type blower is basically a drum, with four fixed fins or vanes which rotate inside the drum. The vanes themselves operate in another drum which, by means of an eccentric cam, revolves in a concentric manner. The changing air volume, from side to side, gives this type blower its 'push'. This blower is efficient but consumes quantities of oil in its operation. Some of this oil gets into the engine through the carburetor. One of the problems of the Zoller has been to keep the amount of oil entering the engine to a minimum.

Though there are variations on the Zoller in Europe the differences are too slight to mention at this time.

The third type, and the most common in the United States, is the centrifugal or impeller type. This type, most commonly used in aircraft, was the same general type as used on Auburn, Cord, Duesenburg and Graham stock cars in this country within the last fifteen years. The McCulloch supercharger made for installation on Ford V8s and Mercuries was of this type. This type blower has a fan or impeller which rotates within a case. This impeller has many curved collector vanes. The air is drawn in at the center of the impeller and thrust with increased force toward the outer edge of the case as the impeller revolves at high speed. In the case of the Duesenberg the impeller was made of forged heat-treated duralumin. It rotated at

six times the engine speed and delivered five pounds of pressure at 4,000 r.p.m. It is interesting to note that the addition of a supercharger to the 265 Horsepower Duesenberg engine increased the Horsepower to 320. The manufacturer claimed that this model car could be accelerated from a standing start to one hundred miles an hour in twenty seconds. A phaeton was driven 129 miles per hour in high gear and 104 miles per hour in second gear.

As the centrifugal type blower has to be driven at speeds higher than that of the engine it takes time to build up the required pressure, thus this particular type has not been used extensively for short track racing.

RADIATORS

The radiator of a car plays a most important part in cooling the motor. There are many types and designs of radiators, all basically the same. One of the popular ones among race car builders is the Harrison 'cartridge'. This has a 'core' which is four inches thick. In addition to the radiator the water tank should be made as large as possible, but should not in any way interfere with the operation of the cooling effects of the radiator. Nothing should interfere with the passage of air through the radiator core. If a four inch Harrison 'cartridge' core is unobtainable, or it is felt that it is too expensive a similar thick-sectioned core can be constructed by obtaining a regular radiator core from a wrecker's, cutting it lengthwise down the center, then putting one half behind the other half and making the proper connections. A new tank will have to be fabricated to match the new thick core.

The tanks should be made of light sheet steel welded, and not of copper with soldered joints. The vibration which a racing car receives would cause the soldered joints to break open causing serious leaks. No car can travel long at high speeds without the proper cooling, and to do so would cause a motor to seize up quickly, and do additional costly damage. In purchasing a used radiator core care should be taken to run sufficient radiator cleaner through the radiator and then flush it thoroughly and carefully. This is a very important precaution.

SPARK PLUGS

There are many types of spark plugs on the market. The engine performance of a racing car greatly depends on the type spark plug used.

The problem in choosing the proper plug for your type engine is to obtain a spark plug which has the proper heat characteristics. There are 'hot' plugs, and 'cold' plugs, and the type plug most suitable for your engine depends on the degree of temperature the plugs attain when in operation. If a spark plug is too hot for your motor it will cause pre-ignition, loss of power, and over-heating of the engine. If a spark plug is too cold it will cause the plugs to foul, and will not give the proper ignition. You can determine whether a plug is too hot or too cold for your motor by observing the appearance of the porcelain insulation. If it is too hot the porcelain will appear blistered, and will have a purple or grey color.

Spark plugs also indicate the proper fuel mixture. Too lean a mixture will cause a white or light brown appearing plug. Too rich a mixture will blacken the plug, and make it sooty looking. When the engine is receiving the correct mixture of fuel the spark plugs will appear to be free from soot, and the porcelain will appear to be colored and dark brown.

CON RODS-BEARINGS-LUBRICATION

Two types of connecting rods are in general use: the drop-forged I-beam type, and the tubular type.

The Model A or B Ford rod is of the I-beam type and is widely used. One of the troubles with this type is that the increased compression ratio and high r.p.m. of the motor causes a strain which 'springs' them out of shape. This 'springing' action is actually a bending which would not even be discernible to the naked eye. However, the constant strain of such a bending action will cause the rod to snap in time. The way to avoid the possibility of a broken rod is to change rods after every few races. Obviously a rod which breaks at speed can cause considerable damage. Any rods obtained for building up a motor should be carefully checked. A 'true' rod not only increases the smoothness of an engine, thus increasing its efficiency, but minimizes the possibility of the babbitt in the bearings cracking.

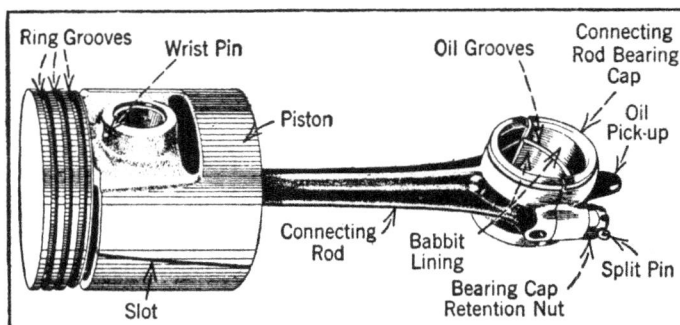

FIG. NO. 10. MODEL A FORD PISTON AND CONNECTING ROD.

The other type of connecting rod most widely used in racing engines is the tubular rod. This is the most dependable rod, but its cost is considerably more than that of the I-beam type. Naturally, even this type of rod will give poor service if improperly fitted, poorly lubricated, or installed without proper inspection.

The lubrication of connecting rods and rod bearings is an important phase of engine building. Insufficient lubrication can cause major trouble. A pressure oiling system is the only method by which proper lubrication can be assured at high engine speeds. With this type of system oil is forced under pressure through a hollow crankshaft to the inside portion of the connecting rod bearings. Care must be taken when fitting the main bearings in order to eliminate the chance of losing any oil pressure. Enough oil pressure is required to assure a constant stream of oil to the con rod bearings. The Model B Ford crankshaft is drilled for a pressure oiling system by drilling a diagonal hole through the cheek of the crankshaft from the center of the connecting rod shoulder to the center of the adjacent main bearing shoulder, thus enabling oil to flow through the crankshaft from the 'mains' to the 'big ends'. (See Fig. 11)

This is a cheap method of oiling the bearings, but it is not too efficient because there is the possibility that small particles of grit or dirt will be forced from the mains to the rod bearing as the oil flows through these small orifices, thus causing additional wear.

The method of drilling the crankshaft used by the builders of specially engineered racing engines is to drill a hole through the center of the crank-pin parallel to the axis of the crankshaft. Then radial holes are drilled through the crank-arms to connect with the hole through the center of the main bearing. A radial hole is then drilled through the crankpin starting from the back and ending with the center hole. A radial hole is then drilled through each main to connect with the oil supply. (See Fig. 11)

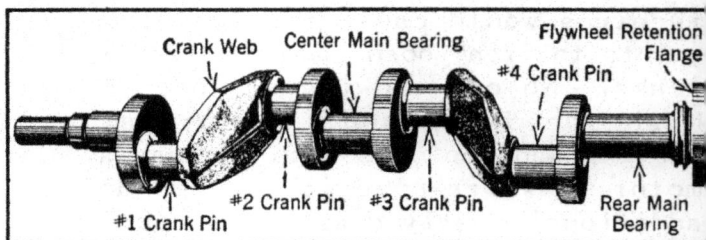

FIG. NO. 11. FORD MODEL A CRANKSHAFT.

The open longitudinal holes on the outer ends of the mains, crank-arms, and crank-pins, are then plugged by tapping and inserting Allen-head set screws. This facilitates removal to clean out any accumulated dirt in these passages due to the centrifugal force for the crankshaft. Dirt packs very hard in these passages and a drill will be needed to clean them out. Though it consumes time and care this should be done after each race meet to assure proper lubrication.

It cannot be stressed too strongly that special attention to good lubrication and the proper installation of connecting rods will pay off in the final analysis, for when a rod cuts loose it is often that extensive damage is done to an engine.

BODIES

Bodies currently being used on both big cars and midgets are of two types: steel, and aluminum. Each type has its advantages and its disadvantages, so the personal choice of the builder is, in the long run, the deciding factor. The aluminum body is easier to work with and considerably lighter than a steel body. In the event of damage to the body it is much easier to repair. The disadvantage of the aluminum body is that after a period of use the constant vibration causes the metal to crack. This is due to the fact that vibration seems to make the aluminum brittle or hard. When this condition arises it is known as 'dead aluminum'. 'Dead aluminum' is much harder to weld than new, or 'live' aluminum. For

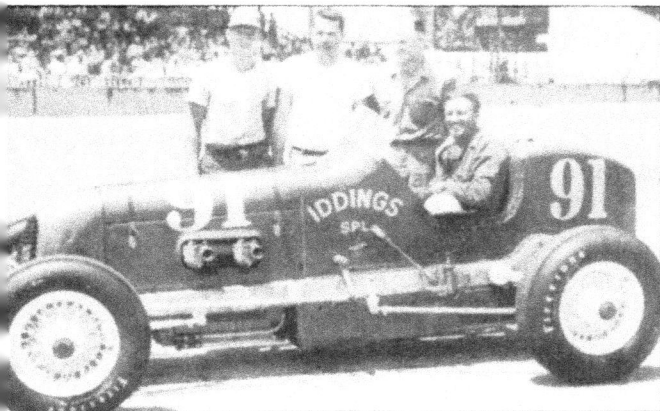

FIG. NO. 13. THREE CARS WHICH RAN AT INDIANAPOLIS IN 1948 WERE ACTUALLY DIRT-TRACK CARS. NOS. 35 (DRIVEN BY MACK HELLINGS) AND 63 DRIVEN BY HAL COLE) WERE KURTIS-KRAFT CHASSIS AND BODIES POWERED BY MEYER-DRAKE (FORMERLY OFFENHAUSER) ENGINES. NO. 91 (DRIVEN BY LEE WALLARD) IS A DREYER CHASSIS AND BODY WITH MEYER-DRAKE POWER.

FIG. NO. 14.   THE FRONTAL ASPECT OF THE "OLSON OFFIE", ANOTHER DIRT-TRACK RACING CAR.   THIS CAR HAS INDEPENDENT REAR SUSPENSION.

those who have had no experience welding aluminum it might be suggested that they obtain some scrap aluminum to 'practice' on before attempting any important body work. Aluminum is definitely tricky stuff to work with and experience is a requisite that cannot be circumvented (See Figs. 13, 14 & 15)

It is advisable that the builder attempt to obtain a tank of Hydrogen, and the necessary Hydrogen gauges. Experience will prove that Hydrogen is the easiest way to weld aluminum as it gives a clean flame.

When selecting sheet stock for the body 14 gauge, or .064 in. is the acceptable thickness. Thicker gauge may be selected if there is much stretching to be done. Quite a few car builders use '2S4', although some use '2SO', which is softer. These figures refer to temper and hardness. If two pieces of metal are to be fastened together, especially a piece of steel, it is advisable to apply a coat of primer between the two. Primer is applied to prevent corrosion which would cause the two parts to loosen. When a hole is drilled or any part of the aluminum is plated it should be reinforced with a backing plate. Either steel or aluminum rivets can be used. This backing plate is to prevent a crack from starting at the rivet hole.

A steel body is much lower in cost and much easier to weld, and steel does not have the tendency to crack from vibration as does aluminum, but it is considerably heavier, and even if you were to use a lighter gauge metal the added weight would be noticeable. As has been stated elsewhere in this book, weight, especially in the midget jobs, is a very important factor, and the elimination of excess weight is of prime importance.

PAINTING THE CAR

A beautiful paint job adds much to the appearance of a racing car. The only satisfactory method of painting a car is by spraying. First it is imperative that no waves, bumps, file marks, or rough spots can be observed on the surface to be painted. The surface should then be thoroughly cleaned with gasoline or thinner and allowed to dry. Then spray on a light coat of primer and allow this to dry. The surface should then be rubbed with one of the finest grades of sandpaper or emery cloth, or with very fine steel wool. If any weld marks or scratches still show they can be smoothed over with glazing putty. After this is applied, and has dried hard, it should be rubbed down smooth. Then add another coat of primer, and rub down smooth again with fine sandpaper and water, always keeping the surface thoroughly clean and dry before applying the next coat.

You are now ready to spray on the first coat of 'color'. Again it is a matter of personal choice whether you use Lacquer or Synthetic Enamel. After selecting one of the two types of paint, in the color that you desire, the first coat can be sprayed on lightly. At least two and preferably three coats should be sprayed on in order to make a good-looking job, keeping in mind that each coat should be allowed to dry thoroughly before the application of the next coat. Between each coat the sand-paper and water 'rub-down' should be given to the entire car. Before the last coat of color is applied the car should be thoroughly cleaned again so that no fine particles, or paint dust from the sand-papering job, remain on the surface. When the final coat is applied slightly more thinner should be added to the

Billy Cantrell

Frank Armi

Danny Oakes

Russ Fields

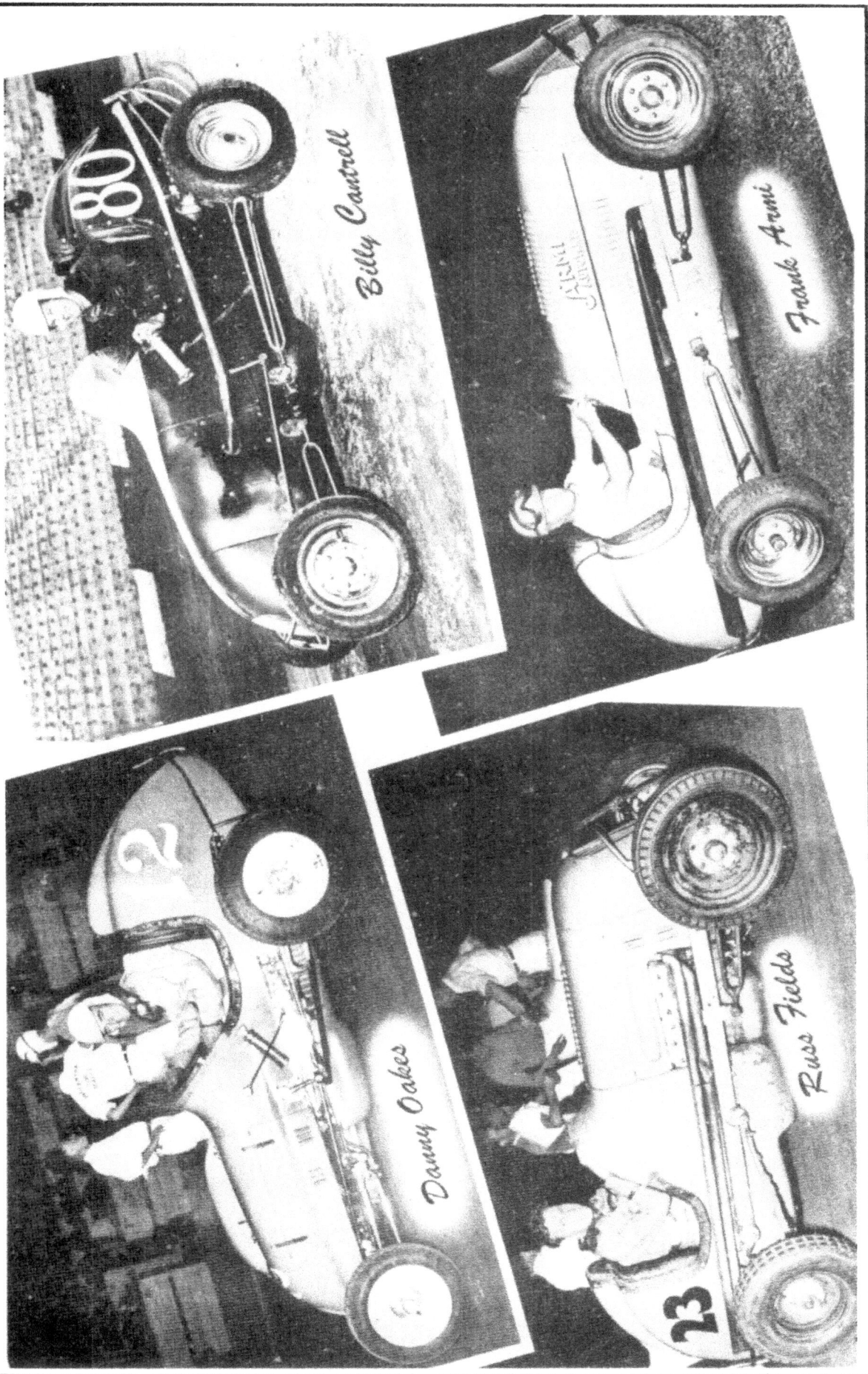

FIG. NO. 15. EXAMPLES OF MODERN BODY WORK AMONG MIDGET RACE CAR BUILDERS. THOUGH THEY APPEAR SIMILAR, INSPECTION WILL REVEAL THAT EACH BODY IS SLIGHTLY DIFFERENT.

color, being very cautious not to allow the paint to run. Allow one day drying time, then rub the entire car down with rubbing compound. After this wax the car with Mac's It, Simoniz, or some other good wax polish, to protect the finish. NOTE: When using synthetic enamel only synthetic thinner should be used. If Lacquer or Duco is used then lacquer thinner must be used. Lacquer and synthetic enamel are basically different and will not mix. The thinners used for each are also different. If lacquer thinner is used on enamel it will cause it to blister, curl up and drop off, almost immediately.

Primer is usually sold in two basic tones, a light gray, and a brown or maroon. If you have selected a light color as the final color of your car the gray, or light primer, should be used. If you choose a dark color, then the reddish-brown primer should be used. Obviously if you use a dark primer and a light color-coat then several additional coats of color will have to be applied to cover the tone of the primer, thus it is easier to use the lighter primer in this instance.

THE RADIATOR SHELL

A neat well-made professional-looking radiator shell adds much to the appearance of a racing car. It is quite a job to hammer out a shell from a piece of sheet metal even for an experienced man. However, there is a way that a satisfactory shell can be fabricated if you have had a little sheet metal experience and can weld. An attractive shell for either a midget racer or a dirt track job can be assembled from the two top rear corners of an old model automobile body. (Fig. No. 16) These corner sections must have the proper contour. If the two corners are cut out completely, and then butted together and welded longitudinally (at dotted lines A and B), the basic shell will be formed. It will then be necessary to cut out

the opening for the grill. This can be circular, oval, heart-shaped, or almost any selected design. The grill itself can be obtained from a wrecking yard, having been selected because it would best fit the style and size of the shell, or it can be cut from one of the many styles of stamped metal screen which are obtainable at large metal supply houses. If the latter type of grill is used it should be selected with sufficient openings for rapid and plentiful air passage. Many of these die-cut screens are attractive, but allow insufficient air to pass through to the radiator core. Obviously the grill opening itself should be of sufficient size to allow a generous quantity of air to reach the radiator. Some drivers weld supplementary screens to the frame horns, in front of the regular grill, to protect the grill from the flying dirt thrown up by cars driving ahead.

GAS TANKS

Many drivers, rather than fabricate a new gas tank, or bother with cutting and rewelding a tank taken from a passenger car, (sometimes a dangerous operation because of the gas fumes which stay in such tanks for many months despite repeated cleanings), prefer to make gas tanks from the five-gallon heavy gauge oil cans such as bulk motor oil is sold in. These tanks are usually of sufficient strength to stand considerable buffeting, such as would be met in the case of a midget car. Because of their capacity limitations however, they would be suitable only for the midgets. A small air vent (to effect rapid filling) should be soldered or brazed to the top, at the opposite side of the regular cap. An extension can be run from the cap opening to the opening provided in the 'fish-tail' of the car. A standard brass fuel line fitting should be brazed to the bottom of the tank so that the line can be hooked up at this point. NOTE: If

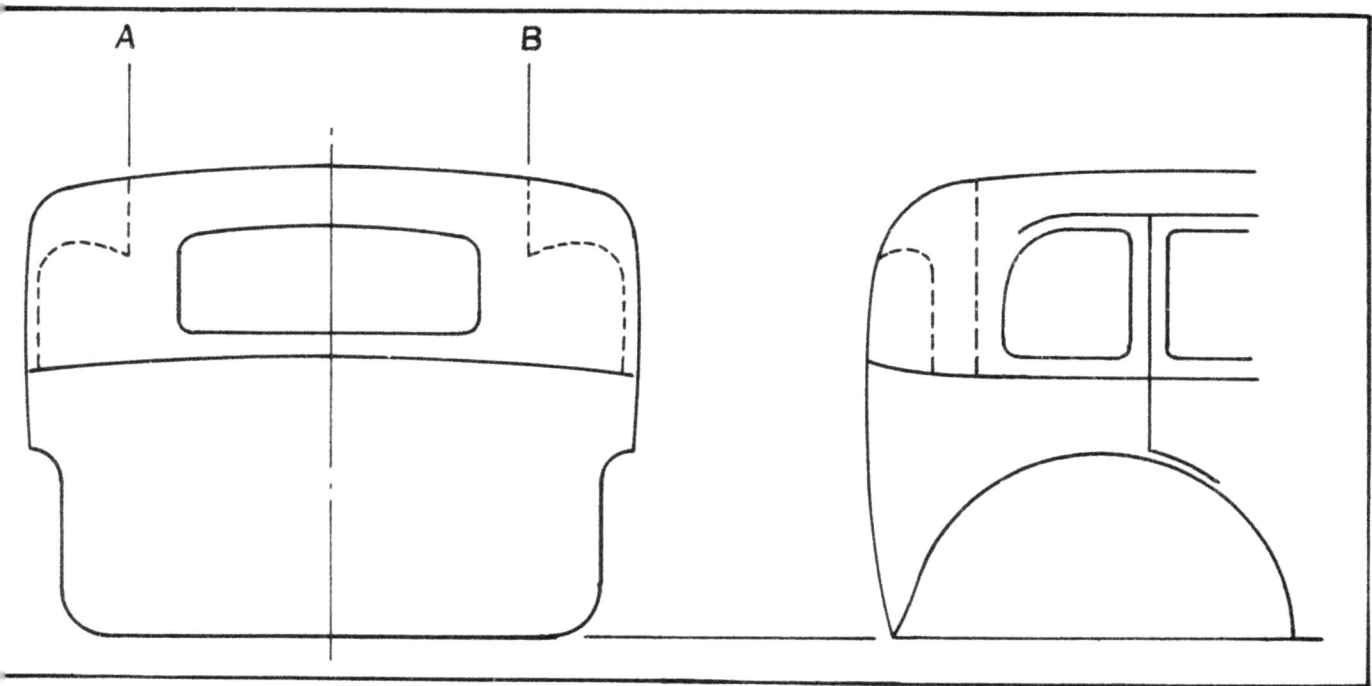

FIG. NO. 16. DOTTED LINES INDICATE WHERE METAL SHOULD BE CUT IN MAKING RADIATOR SHELL FROM OLD AUTO BODY.

FIG. NO. 17. DISC BRAKES INSTALLED ON HAL COLE'S DIRT TRACK RACING CAR. LEFT: FRONT WHEEL, SHOWING HEAT DISSIPATING FINS AND VENTILATION LOUVRES. RIGHT: REAR WHEEL BRAKE WITH SCOOP-LIKE AIR INTAKES FOR COOLING PURPOSES. NOTE SPLINED "KNOCK-OFF" HUB FOR RAPID WHEEL CHANGES.

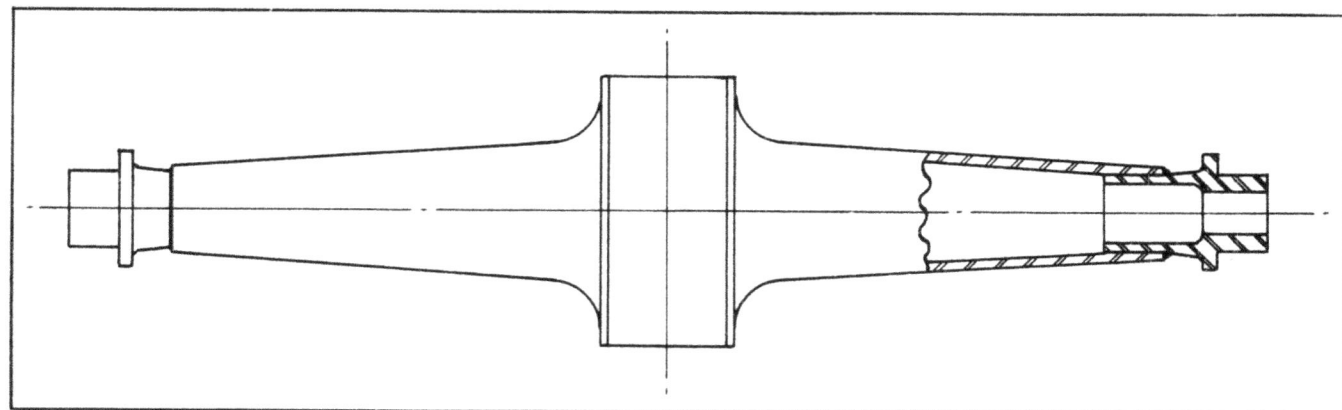

FIG. NO. 18. METHOD OF CUTTING FORD REAR AXLE HOUSING AND RE-WELDING TO OBTAIN NARROWER TREAD.

the fuel is fed to the carburetor by means of air pressure in the tank (actuated by a hand-operated air pump), an air vent cannot be installed, as no pressure could be built up. And, by the same token, the filler cap will have to be an air-tight fit if this system is used. However, if either a mechanical (diaphragm) or an electrical (Auto-pulse) fuel pump is used the air vent is permissible, and the filler cap need be only tight enough to secure.

The tank, of the type described, can be welded to the rear frame members, or a supporting frame built around it in such a manner that it can be bolted in and tightened in position, but still will be removable, if any repairs are necessary on the tail of the car. However, it must be absolutely secure as from three to five gallons of fuel constitutes enough weight in the rear end of the car as to cause considerable inertia when cornering at high speed, and at that point it is not prudent to have the fuel supply coming loose from the frame.

So-called 'Jeep' auxiliary tanks, such as can be purchased at War Surplus stores, can also be used for gas tanks, as they are of sufficiently heavy gauge metal to withstand rugged treatment, but here again is the problem of brazing metal which has contained a highly volatile fluid. There have been instances where repairs being made to gas tanks (by welding) caused serious explosions, so welding to any tank which has contained gasoline is not advisable.

CHASSIS COMPONENTS OF MIDGET AND DIRT TRACK RACING CARS

1. The Frame

Choice of a suitable frame is up to the builder of the car. Whether the car is to be a Midget, or Dirt Track (3/4) car is a determining factor in the type frame to be selected. Midget frames can be built up from Model T (Ford) side rails (channel), Model A (Ford) side rails, aluminum channel sections, or tubular steel. The larger cars can be built from channel-section side rails from several passenger cars (obtainable from wreckers), or from aluminum channel, depending on the weight requirements of the car. One of the determining factors in the type side rails to be chosen is the type of springing and suspension to be used. Dirt track cars almost all use transverse springing (similar to Ford Models T, A, B, & V8) at the rear of the car. Some of these cars also use transverse springs at the front end, others use semi-eliptic springs of a very flat arch. Up till recently most of the midgets used transverse springs at both front and rear, but torsion bar suspension is increasing in popularity, as are variations of rubber block-mounted suspension. If these latter types are to be employed in the building of a midget racer, then the frame can be fabricated from light but strong steel tubing. If the more conventional transverse leaf system is used, then the equally conventional channel side rails are usually employed. These side rails can be held together (transversely) by steel tube cross-pieces securely welded in place. Designs and actual measurements for both midgets and dirt track cars vary with the power plants to be fitted, and the wheelbase requirements. Suffice to say that the side rails on both these types are absolutely parallel and have no taper such as is found in all passenger cars. The amount of taper to the front frame horns, or whether any will be needed at all, and the amount of 'kick-up' at the rear, are both features which are flexible, and are determined by the suspension systems selected.

2. The Front Axle

The ease of handling and stability of a racing car, whether midget or dirt track, is in direct proportion to the soundness of design, proper lining up, and correct installation of the front axle and its steering components. Some cars are easy to handle and require little effort when going into or coming out of turns, others must be 'fought' most of the time, still others are 'floaters'.

One of the items most directly related to the flexibility of a steering system is the degree of caster of a front axle. Increasing the degree of caster will make the car's steering more stable, but at the same time will make it more difficult to steer when cornering. The more the caster is increased the more the car wants to follow a straight path. 'Caster', as the word denotes, is the degree of inclination off the true vertical, to which an axle is tipped (on its lateral axis). Thus, if you were to sight lengthwise across an axle from the left spindle to the right spindle, you would notice that both king-pins (right and left) were tipped so that the top of the king-pins were slightly forward of dead center of the axle, and the lower part slightly to the rear of the center-line of the axle. This angle is almost imperceptible, but is very important. It is this slight inclination which makes the wheels 'follow', just like the caster on a chair.

Another item directly connected with the ease of handling of a car is the 'camber'. As with 'caster' this is a most important factor. All front wheels (when observed from the front of the car looking toward the radiator) are slightly tipped on their vertical axis, so that the points where the tread of the tire touches the ground is slightly less than at the tops of the wheels. This camber is greater in some cars than in others. It

was most easily noticed on the Bugatti cars, both passenger and racing. Almost all models of Bugatti had a very pronounced and evident camber. One system of determining the proper amount of camber is to draw an imaginary line through the dead center of the king-pin. A projection of this line should hit the ground at dead center of the tread of the tire.

Thirdly, an important steering factor is 'toe-in'. All cars have a certain amount of 'toe-in'. This means that both wheels are not absolutely parallel, but toe into each other very slightly at the front. Thus a measurement taken from the tread centers, or at the inside edges of the rims) would be slightly less at the front of the wheels than one taken at the same point on the rear of the wheels. This toe-in should never be more than 1/16th of an inch, for good handling.

One of the cheapest axles to convert to racing car use is the Ford Model A or V8 axle. It makes an excellent front end for a transverse-sprung chassis, and needs very few alterations. It is necessary to reinforce the spindles with 3/16th inch steel plate bolted on in place of the brake drum backing plate. Another method is to cut down the front brake backing plate itself. The steering arms will have to be bent slightly so that the tie rod will clear the radius rods when turning at full lock position.

If a tubular axle is desired the Chrysler 70, or the Franklin axle have been very popular. However, if these are used with transverse springing some machine work will have to be done in order to make the proper 'perches' for the spring shackles. Plates should be welded to the axles and holes drilled in these plates. Care must be taken to give the axle the correct amount of caster. Many other types of rebuilt axles are in general use, but the above-mentioned are the most popular.

Front axles for midget racing cars are constructed in practically the same way as for 'three-quarter' cars except that they are cut and re-welded to accommodate the narrower tread. The most common in general use are the Ford Model T, Model A, and V8. A small section is generally cut out of the center and then both ends welded together, sometimes with additional reinforcement at the welded joint. The front spring (transverse) on a midget racer measures 18½ inches from shackle to shackle, so the spring will have to be rebuilt to fit these dimensions. The degree of caster under loaded conditions varies from 3° to 6° so the builder will have to determine what degree of caster he wishes to operate with.

The tubular axles being used on midget racers are generally specially made of chrome molybdenum steel tubing with special spindles, or with stock car spindles that have been altered. Practically all midget race cars use either Ford or Willys spindles. Brakes and drums are usually special. Four-wheel brakes are increasing in popularity among the midget car owners, and on many of the cars that are equipped with two-wheel brakes (on the rear wheels) the brakes are actuated by a hydraulic system. Quite a few of the track roadsters and dirt track cars are now being equipped with hydraulically operated disc brakes. Disc brakes proved to be very efficient on planes during the war, and are working out quite well on racing cars. (See Fig. 17)

3. The Rear Axle and Differential

There is practically no difference whatever in the construction of the rear end for either a midget or a three-quarter car. The major difference being that the midget has a narrower tread. There are two types of rear ends in use on both the big cars and the midgets. One is the semi-stock type,

the other is the 'quick-change' type. The semi-stock type, converted from a passenger car rear end, (usually Ford A or B is always locked, for both midget and big car use. In this manner both rear wheels are in positive drive at all times, giving full traction-- thus differing from the passenger car differential which allows one wheel to turn faster than the other when making a relatively tight turn, such as on a city street. In a passenger car one wheel is doing practically all the pulling in a short turn. (When your car is on a lube rack and the motor is idling you can hold one wheel while the other will continue to turn). In a racing car with a 'locked' rear end both wheels turn at the same rate all the time, giving the affect of a solid rear axle. One wheel necessarily must skid some, but in most cases (much more so in the case of the midgets) both wheels are sliding in the turns anyway as the car has deliberately been put into a 'broadslide' by the driver. Consequently both wheels, though skidding, are doing an equal amount of skidding, and are under closer and more direct control of the driver. It has been said that a good driver (racing in the turns) does as much steering with his foot (the throttle) as with his hands. It is certainly true that on high compression cars the angle and amount of skid is under close control, and can be varied at.

A rear end, which is being converted from stock (passenger car) to track use, and is being 'locked', can be locked by any one of several methods. One method is to add another spider gear in the differential then pour babbitt into the gears to eliminate any backlash. This is hard both on the gears and on the axles themselves. The more efficient, but more expensive method, is to lock both axles together by means of a splined steel spool specially machined to fit over the inner ends of both the

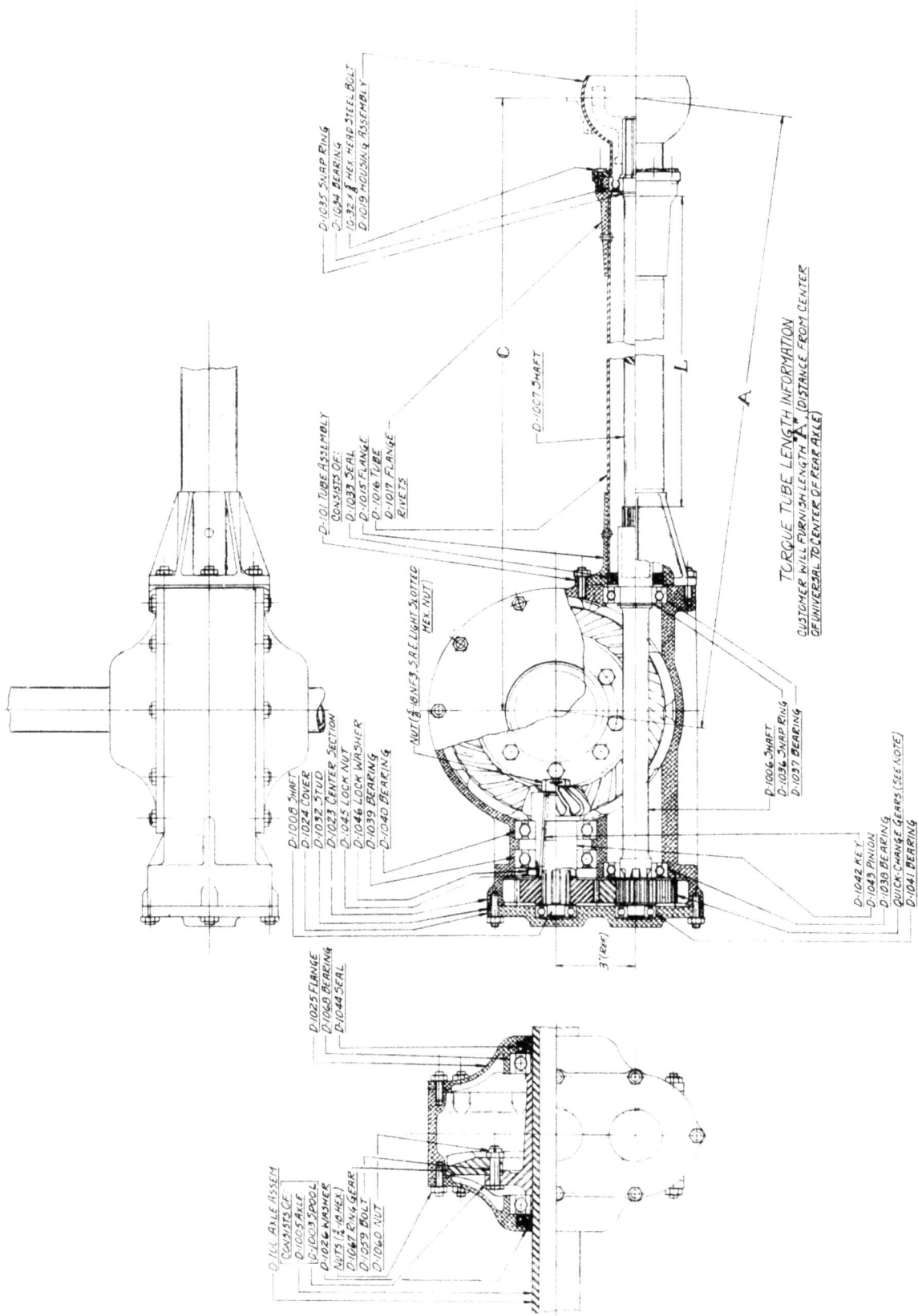

FIG. NO. 19. THREE-WAY DRAWING OF A "QUICK-CHANGE" REAR END CLEARLY SHOWING HOW DRIVE SHAFT RUNS UNDER REAR AXLE TO SPUR GEARS, THENCE TO RING GEAR VIA REAR-MOUNTED PINION. THE TWO SPUR GEARS CAN BE SPEEDILY CHANGED TO AFFORD AS MANY AS FIFTEEN DIFFERENT RATIOS. WITH THIS TYPE REAR END THE AXLE SHAFTS ARE OPEN, AS THE HOUSING DOES NOT EXTEND OUTWARD TO THE BRAKE DRUMS.

right and left axle. Heavier axles should be installed when these changes are being made as all the power of the engine, and all the driving force, is put on the outer axles. An axle with the splined spool lock, and with safety hubs will be lighter than 'stock' and trouble-free.

When reconstructing a Model T or Model A rear axle housing for Midget use it can either be cut (hacksaw or torch) and rewelded to the narrower tread measurements, or the rivets can be cut out, the brake housing slid in (toward the center), rewelded in position, and the excess stock cut off. (See Fig. No. 18). In the instance of the Model A rear end the Emergency Brake carrier and the regular rear brake drum should be retained, with the service brake removed on a lathe. In rewelding the rear housing it is important to have removed the grease retainer previously. It must be re-installed, in line, after the welding.

In both Model T and A rear ends the drive shaft housing (torque tube) should be cut, shortened, and re-welded back near the differential housing. Keeping the tube lined up properly when re-welding is the most important item. The drive line also will have to be cut and shortened. It is advisable to have a machine shop do this work as it will have to be splined at the universal--joint-end, or tapered and made to fit the pinion gear at the rear end. Though it is cheaper to cut the drive line in the middle and re-weld it, or slip a tube over it and weld or rivet the tube, this procedure is not at all advisable because it is seldom that the shaft lines up properly after such an alteration, and any variation from an absolutely straight line will cause 'whip-lash' causing the drive shaft to break. There is an enormous amount of strain on the drive shaft to break. There is an enormous amount of strain on the drive-shaft of a racing car.

The other type of rear end, and major difference from the converted stock rear end, is the 'quick-change' type. (See Figs. 19 & 20). This type carries the drive line under the rear-end housing (as in a hypoid drive, or worm drive), then through the use of two spur gears is geared to the ring and pinion. This set-up has a two-fold advantage; it drops the entire drive-line low, so that the driver can sit low in the cock-pit; and, this is even more important, it provides a readily accessible housing from which the gears can be removed, and replaced with gears of a different ratio, and 'buttoned up' again within the period of ten minutes. For this reason it is aptly named a 'quick-change' rear end. The changing of ratios is sometimes most important to the winning of events. With this type of rear end it can be accomplished between races. This type of rear end is now being used extensively by dirt track cars and midgets alike. It should be added, however, that these 'quick-change' rear ends are very expensive, though they give about fifteen different possibilities of gearing.

4. Safety Hubs

So-called 'safety' hubs are a universal and very necessary feature of racing car rear axles. They are used on all Indianapolis cars, most dirt track cars, and in increasing numbers on midget cars. Actually, safety hubs are nothing more than 'full floating' axles such as are used on all heavy duty trucks to avoid the loss of a rear wheel if an axle should break. Semi-floating axles are used on all 1949 American passenger cars, and with this design the axle shaft carries the load of the car from the wheel bearing outwards to the wheel hub or bolt flange. Prior to 1949 however, Ford products used a compromise design known as the 3/4 floating type rear end. With either a 1/2 or 3/4 floating axle an axle shaft

BRAKE DRUM

DRIVING
TEETH

WHEEL
PILOT DIAM.

309

AXLE SHAFT

WHEEL
BOLT DIAM.

OIL SEAL

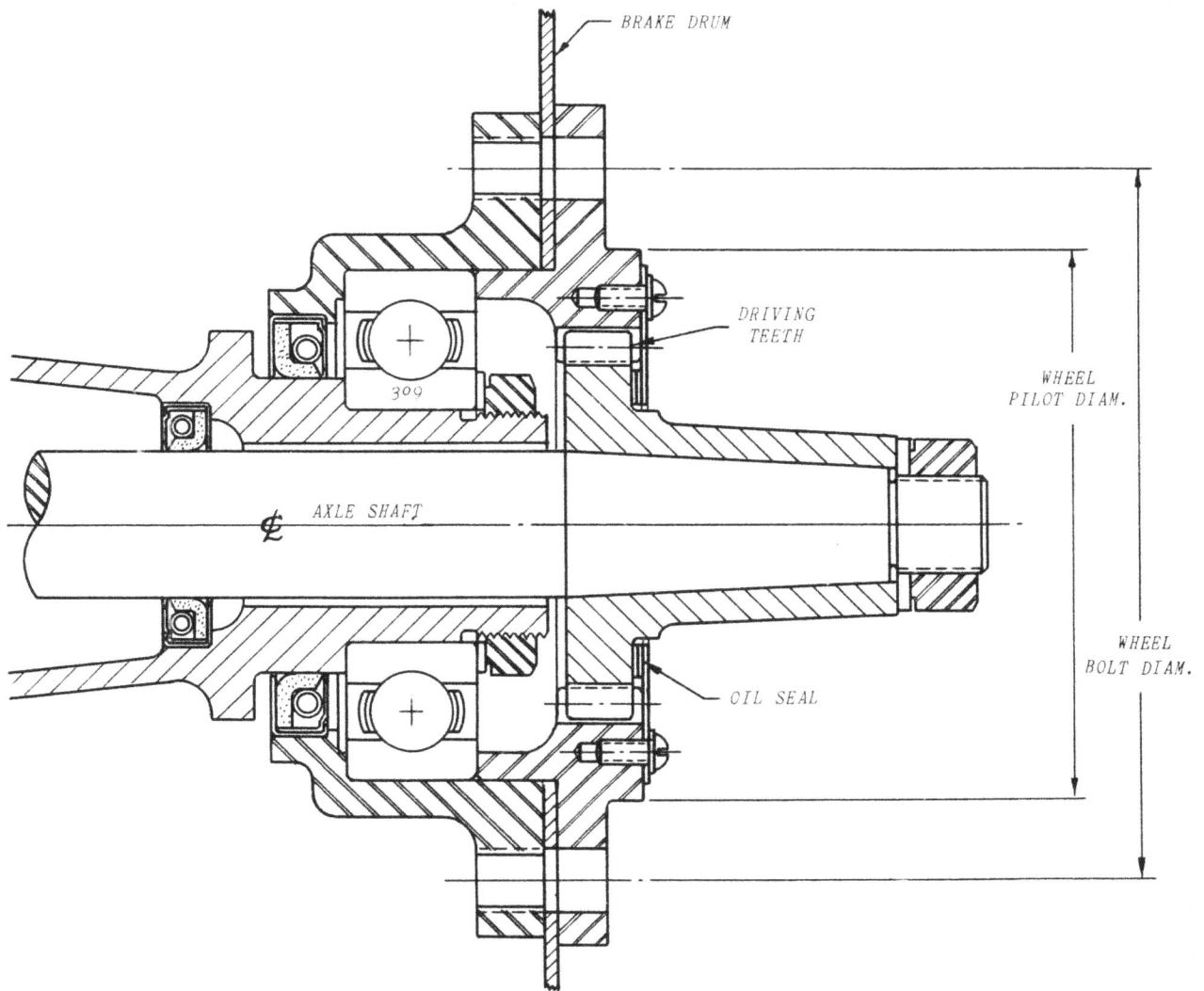

FIG. NO. 21.   DETAILED DRAWING OF METHOD OF CONVERTING STOCK FORD REAR AXLE TO A "SAFETY HUB".

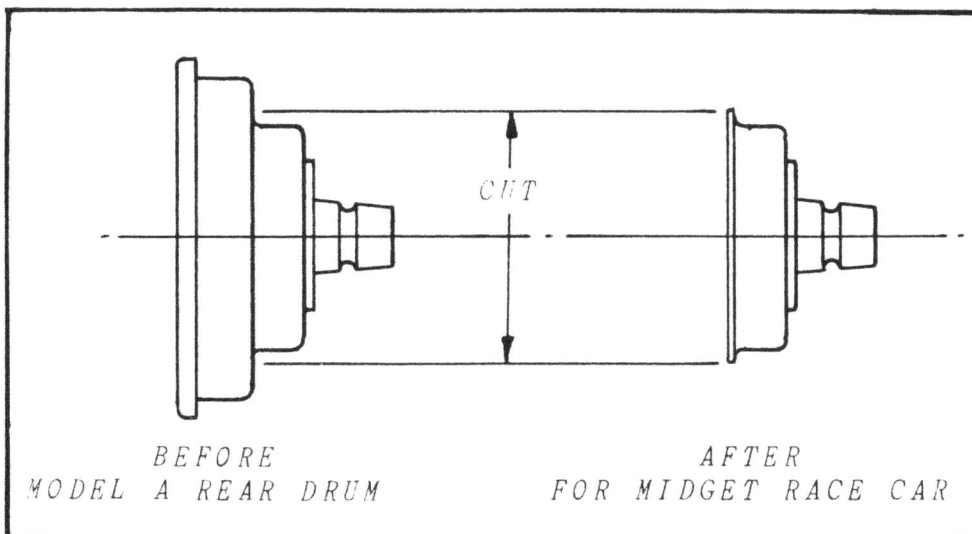

CUT

BEFORE
MODEL A REAR DRUM

AFTER
FOR MIDGET RACE CAR

FIG. NO. 18B. METHOD OF CUTTING MODEL A BRAKE DRUM FOR USE WITH MIDGET HUB.

breakage is hazardous. The Ford rear end, however, is easily converted to the full-floating or 'safety' type.

The illustration (Fig. 21) shows one possible conversion for Ford axles. This particular design shows driving teeth which permit some axial misalignment, and thus eliminates some very difficult machine work in locating the taper bore correctly, which would be necessary if the two members with mating gear teeth were in one piece. a study of truck full-floating axles will show other possibilities, but the design shown is recommended. Trucks employ a pair of opposed Timken bearings for each wheel, but the single row annular ball bearing is more than adequate for a racing car, and the reduced weight and reduced rolling friction are distinct advantages. While the single row bearing is not technically a correct application, (due to combined radial and thrust loads) they seldom require replacement.

For the builder desiring the best, a pair of angular contact bearings (No. 209), or a double row (No. 5209) would be ideal. Norma-Hoffman make a single row bearing, known as a 'duplex angular contact' (No. 345 CD) which is a second-best choice but they are not easy to find. Bearings that are most often used in the construction of a 'safety hub' for a midget car are 'Hess-Bright' No. 308, No. 309, or No. 1309. Regardless of bearing choice it must be securely clamped between shoulders on both outer and inner races. The clamp-nut must be positively locked by a washer having an inside tongue which fits a key-way milled through the male threads, and bent over against the hex flats on the outside of the nut. All bearing manufacturers supply catalogs which give exact dimensions and tolerances for housing bore, shaft diameter, and shoulder diameter, as well as standard clamp nuts and locking washers. A Ford

axle pinion shaft nut may also be used.

The oil seals shown are well worth the extra trouble in building. The outer seal consists of a simple flat plate which clamps several thicknesses of ordinary gasket material, determined by the final positioning of parts after final assembly.

With a hub such as this the wheel will stay in its proper position on the axle housing should the axle shaft fracture, and thus the car can be brought to a safe stop without the loss of a wheel at high speed. Safety hubs often incorporate 'knock-off' hubs, (See Fig. 17) which enable the wheel itself to be changed rapidly by means of a splined hub with a 'spinner' locking ring. These hubs were commonly used with Buffalo, Dayton, or Rudge-Whitworth wheels. One of the last passenger cars to use a similar (though not identical) principle was the Auburn.

The illustration which shows construction of a safety hub can be used as a guide for either a midget car or a larger car, a larger bearing being used with the larger car. During construction care should be taken that all parts are made with close tolerances (or, where possible, press fit) as there should be no play. The brake drum will have to be altered to fit the safety hub. This drum can be fastened to the safety hub by removing the studs from the drum, using the holes of the studs as guides, and installing new bolts to facilitate holding the hub, drum, and wheel together as a unit.

5. Wheels for the Midget Racer

There are several types of wheels being used on the midgets. Disc wheels have become standard and are the most widely used. Cast magnesium wheels are now almost universal as steel wheels are prone to failure due to 'fatigue'. Factory-made wheels can be obtained from racing supply houses and wheel manufacturers.

One 'home-made' type wheel that was used extensively in the earlier days of midget racing, and before speeds got so high, was fabricated from 1927 Model T Ford brake drums. If one wants to make a set of wheels by this method the following directions will apply. These wheels are not recommended where high speed driving is to be undertaken.

The drum should be put into a lathe and cut down to facilitate installation of a standard 12 inch midget drop-center rim. It will be necessary to obtain this rim from a wheel distributor, or parts supply house. The drum and rim must be firmly held during the welding process, so it is advisable to make some sort of jig to hold them in alignment. Weld the rim to the disc (drum) using spots about one inch apart, staggering the spots from one side of the disc to the other. The reason for not making a continuous weld around the inner circumference of the rim is because this is liable to cause warpage of the rim, thus causing the wheel to be untrue, and run with a wobble. This, in turn, would cause the car to handle poorly at speed.

6. Tires

A good set of tires which will suit the track (surface) as well as the car, will add much to the safety and performance of the car. On dirt track cars many types of tires are being used, from passenger car tires to special racing tires.

All the Midget tires are 12.00 rim size, however, so are specially made for the midgets. The general practice, both with the big cars and with the midgets, is to use smaller section tires on the front wheels--these tires being grooved longitudinally, (in the same direction as the tire--fore and aft). Larger section tires are generally used at the rear of the car, and these are grooved laterally (cross-wise), in an effort to obtain the maximum traction, or 'dig' from the driving wheels. The grooving of the front tires in the manner described causes them to track better, so that when steered to one side or the other they tend to hold true to the course intended. Many variations of the two patterns of grooves described are in use, the treads supposedly have advantages of one kind or another, as claimed by the manufacturer. The surface of the track to be driven on might alter the selection of certain treads also. Obviously, if you are using a tread which will allow the wheels to slip and spin too freely much power and speed is going to be wasted. By the same token, a tread which allows no slippage whatever might make a car tough to handle when the driver wishes to broadslide the the car. Some drivers use tires on the rear end which have for their tread merely a lot of oversized knobs. Track conditions, the car's power and handling, and degree of bank to the turns are all factors which govern the selection of the most suitable tire tread. No hard and fast rule can be given here for selecting the proper tread.

For races held on paved asphalt tracks Midget cars use tires known as 'slicks'. These have no discernible tread but are smooth in appearance. The use of such tires enables the drivers to broadslide in the same manner that they would on a dirt track or similar loose surfaced track. The tires are quite broad at the point of contact with the track, thus giving the necessary traction.

Some midget cars have trouble with the rear tires slipping around on the rim of the wheel, tearing out the valve stem. This can be eliminated, or at least minimized, by drilling holes about 2 ins. apart along the edge of the rim and inserting screws directly into the casing of the tire. These screws should not be put in deep enough to penetrate to the tube, but just deep enough to make the casing 'seat' properly, and keep it from turning on the rim. This solution can also be applied to dirt track cars if this sort of trouble is experienced.

7. Balancing a Racing Car

The balance of a racing car is a most important factor in the proper handling of the car. Both Dirt Track cars, Hot Rods, and Midgets should be properly balanced. However, this balancing is most essential in the Midgets, as they are the most sensitive to handle of the three types. Modern day passenger car practice is to place the engine far forward, in many cases directly over the front axle, and in some cases, partially ahead of it. This is very bad practice in the construction of a racing car. Racing car engines, in dirt track or three-quarter cars, should set slightly to the rear of what was formerly the conventional position for an engine. Keeping the weight of the car well centered makes for more stable handling and good cornering. In Midgets, of course, the quarters are close, and there is not much room for shifting an engine backward or forward.

To properly ascertain the balance point of any racing car, either three-quarter size or midget, the car should be lifted on some sort of support placed 'amidships' or at the approximate center of the car (laterally, of course). The car should balance or teeter at a place very near to the center point between the two axles. If the car is badly off balance the engine should be moved forward or backward a sufficient amount to bring it into correct balance. The weight of the engine (in front) is offset by the weight of the driver and the gasoline (in the rear) thus making for an equitable distribution of weight.

8. Radius Rods

Radius rods add much to the stability and handling of a race car for they serve to keep the front and rear axles firmly in a lateral axis. These radius rods are more important on a cross (transverse) spring job than they are on a three spring (rear cross, front semi-elliptics) or a four spring (semi-elliptics all around) chassis, as these two latter types do not depend entirely on radius rods to keep the axles in a rigid plane and parallel.

The front radius rods should be made as long as is feasible. They can be constructed from the 'wishbone' of a Model T, A, or V8 Ford, as they have the proper connections at the front end for fastening to the front axle spring perches. The radius rods could also be made from steel tubing fitted with the proper attachments for hooking to the front axle, and with a ball and socket joint for connecting at the rear (or frame) end. (See Figs. 15, 22 & 23)

In the same manner, the rear radius rods can be made from any of the above-mentioned Ford rear radius rods, or from steel tubing as is desired. In either case the front ends of the rear rods must be fitted with ball joints. These can be the same ball joints as are found on the tie rods of Model A or V8 Fords, as they are complete with grease fittings. The front end of the rear radius rods should be anchored (via this ball joint) to the chassis at a point directly opposite the front universal joint of the drive shaft. (See Fig. No. 22) This is an important factor because the entire rear end system must (when bounding up and down on an uneven track) move in a vertical plane as a unit. Thus the universal, and pivot points of the radius rods must be in a direct line. The radius rods must be made strong as there is considerable strain on them.

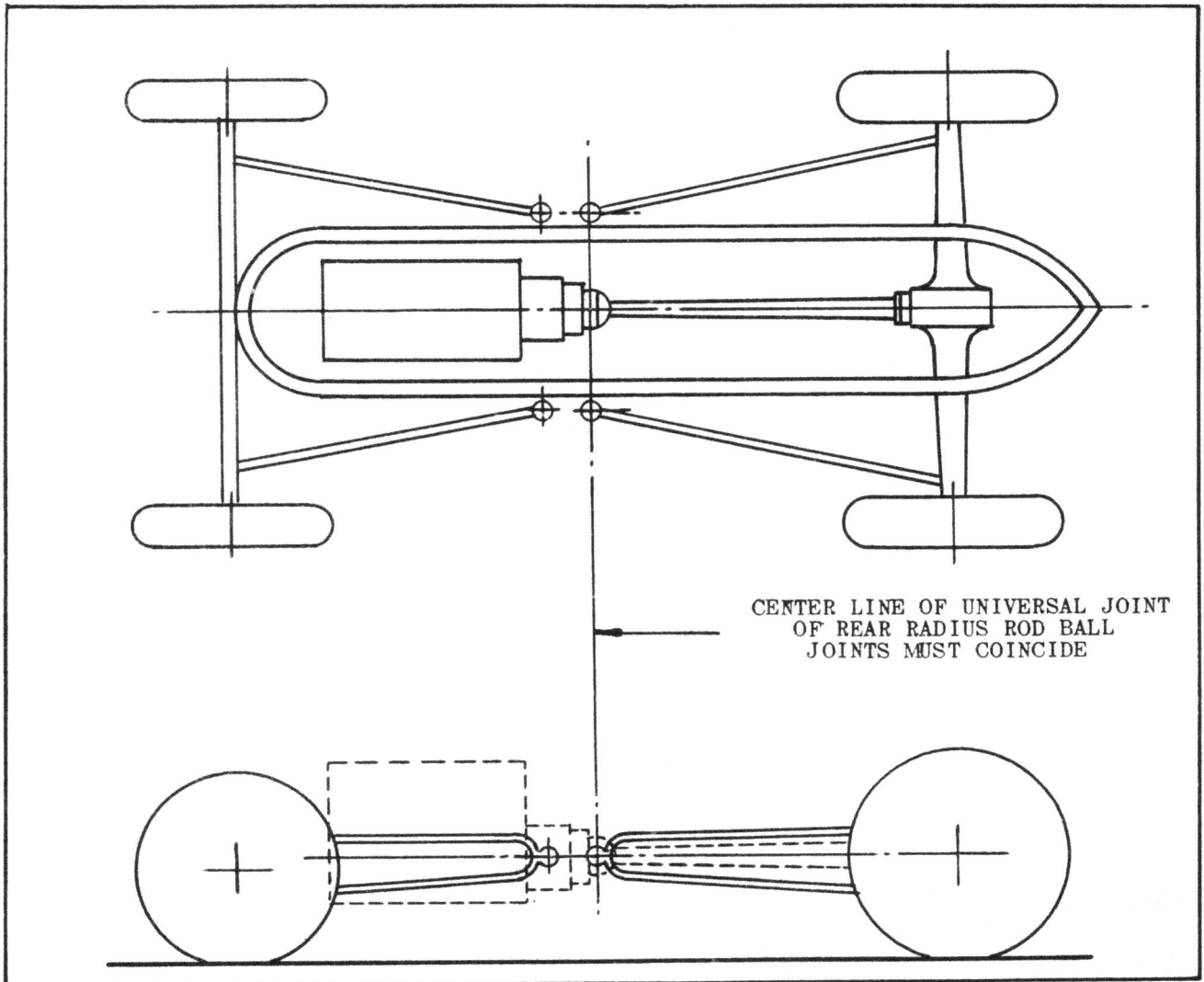

FIG. NO. 22.   DIAGRAM SHOWING HOW PLACEMENT OF UNIVERSAL JOINT DETERMINES
PLACEMENT OF BALL JOINT AT REAR RADIUS RODS.

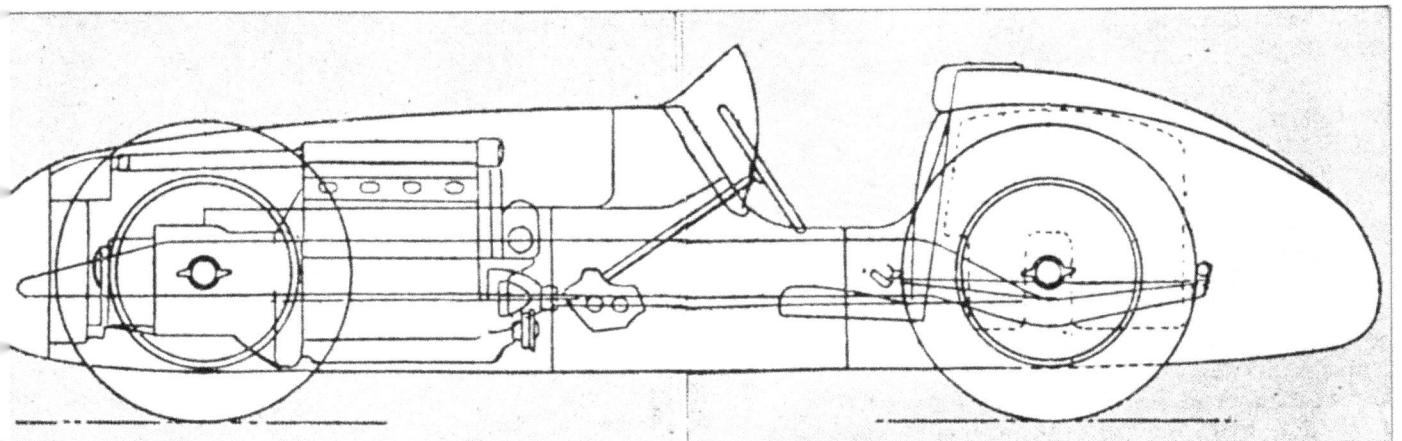

IG. NO. 25.   LINE DRAWING (SCALE 1:25) OF LOU MOORE'S BLUE CROWN SPECIAL. IDENTICAL CARS OF THIS
ESIGN PLACED FIRST AND SECOND IN BOTH THE 1947 AND 1948 INDIANAPOLIS 500 MILE RACE.   THE CARS
RE FRONT-DRIVE, POWERED BY MEYER-DRAKE 4-CYLINDER 274 CU. IN. ENGINES.

35

FIG. NO. 23. CLOSE-UP OF MIDGET CAR SHOWING: (1) HOW RADIUS RODS ARE ANCHORED TO FRAME; (2) HAND BRAKE; (3) PRESSURE PUMP TO GAS TANK; (4) PITMAN ARM OF STEERING ASSEMBLY, AND (5) BUMPER GUARD.

Walt Faulkner

FIG. NO. 20. PHOTO OF MIDGET RACE CAR WITH "QUICK-CHANGE" REAR END.

9. Steering Assembly

One of the most popular and most inexpensive steering gears is that used on the old Franklin automobile. This must be cut and altered to fit your car. A new cross shaft will have to be made which will go through the body at the cowl, and clear the frame. The Franklin gear can easily be adjusted when a little wear has taken place in the gears. It will be necessary to weld a piece of tubing to the gear case to support the whole assembly. Connecting rods which have the same diameter (bearing i.d.) as the tubing can be used to support the assembly--the lower ends being welded to the frame or the entire supporting bracket assembly can be made from steel tubing or angle iron. There are other specially made steering gears on the market, but their cost is considerably greater, and their efficiency no better.

The steering wheel (spokes and center portion) is made from 14 gauge (thickness) 'spring' sheet steel. It can contain either three or four spokes. (See Fig. 24) The narrowness of the spokes depends on the number of spokes, and on the flexibility desired. The rim of the steering wheel can be made from 1/2 inch tubing (for a midget), or larger size tubing for a larger car. This tubing should be electrically welded to the webs or spokes. The rim should then be taped all around. A wooden rim could be substituted if desired by the builder. Likewise, the diameter of the wheel is left up to the builder, as it depends on the clearance in the cockpit, for good freedom of motion in all directions.

BENDING TUBING

One of the cheapest and most efficient methods of bending tubing without the use of expensive equipment is to fill the tube with dry sand, plug both ends, and bend around some stationary solid object. For some bends the tubing will have to be heated at the point of the bend. Some pipe and tubing distributors will bend tubing to any desired radius for a small additional charge. The reason for filling the tubing with sand is to avoid flattening at the point of the bend.

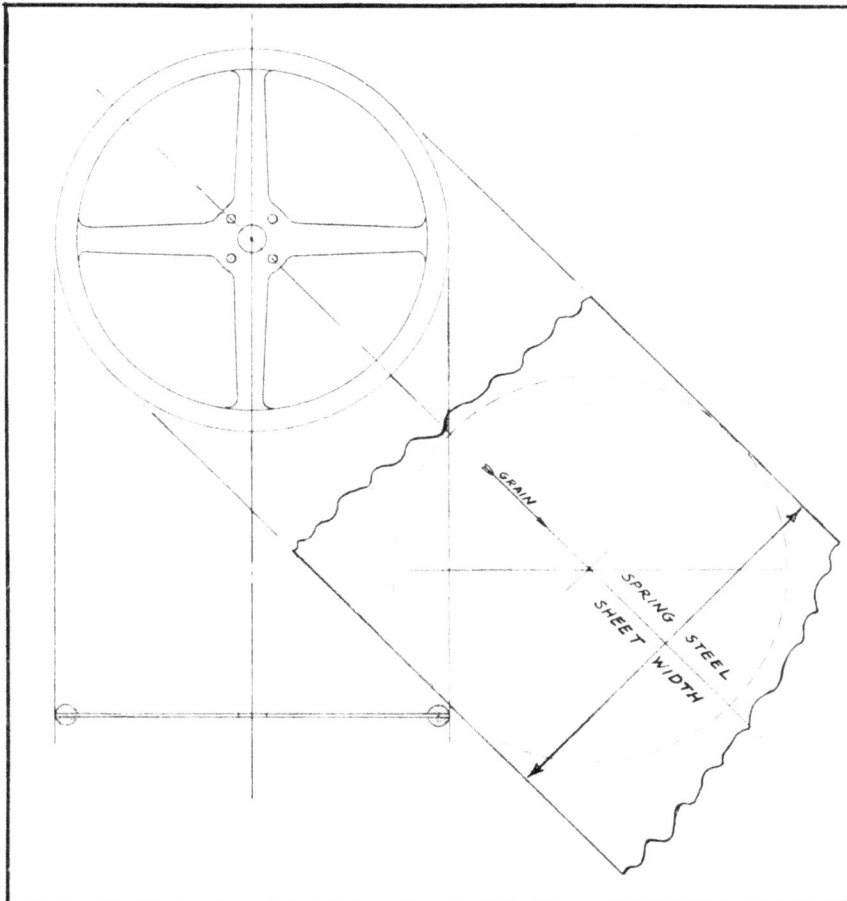

FIG. NO. 24.   DRAWING SHOWING HOW STEERING WHEEL SPIDER IS CUT FROM SHEET STEEL. GRAIN OF STEEL SHOULD RUN DIAGONAL TO SPOKES OF WHEEL AS INDICATED.

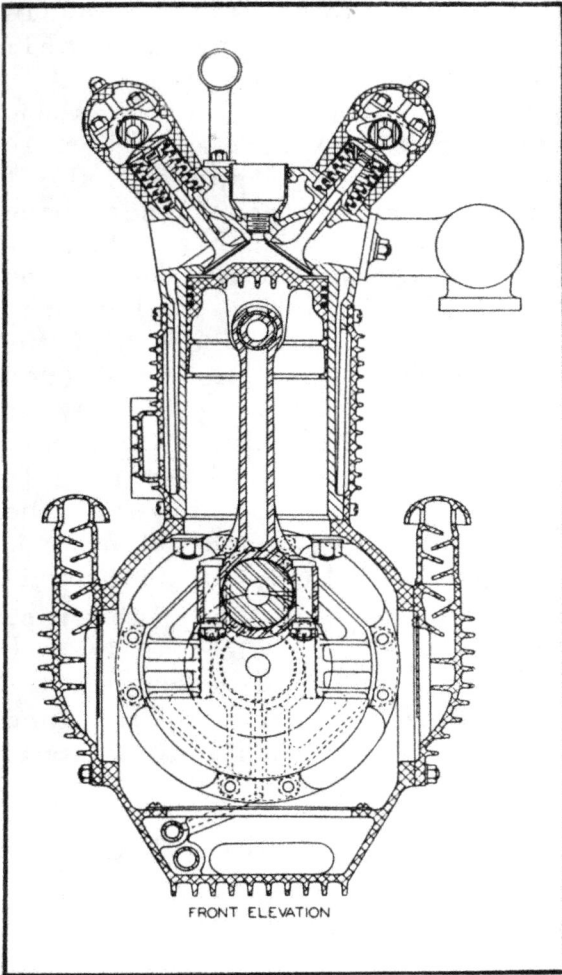

FIG. NO. 26. CROSS-SECTION OF THE MEYER-DRAKE ENGINE USED BY MANY INDIANAPOLIS CARS.

FIG. NO. 27. CLOSE-UP VIEWS OF THE ENGINE OF THE BLUE CROWN SPECIAL SHOWING AIR INTAKES OF THE TWO SIDE-DRAFT RILEY CARBURETORS. NOTE HOW LOW ENGINE SETS RELATIVE TO FRONT WHEEL.

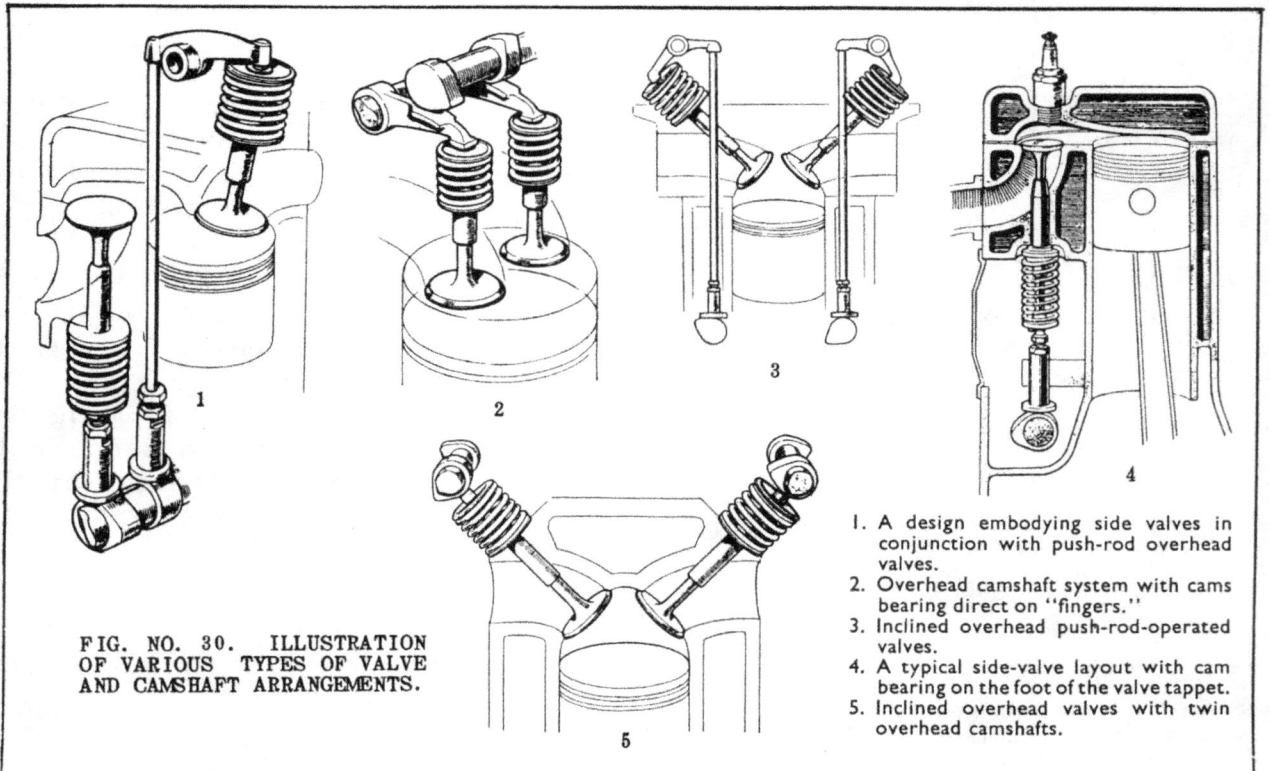

FIG. NO. 30. ILLUSTRATION OF VARIOUS TYPES OF VALVE AND CAMSHAFT ARRANGEMENTS.

1. A design embodying side valves in conjunction with push-rod overhead valves.
2. Overhead camshaft system with cams bearing direct on "fingers."
3. Inclined overhead push-rod-operated valves.
4. A typical side-valve layout with cam bearing on the foot of the valve tappet.
5. Inclined overhead valves with twin overhead camshafts.

38

FINDING PISTON DISPLACEMENT

The Piston Displacement, or 'swept volume', is the volume (of space) within the cylinder between the points where the piston is at its very lowest point (bottom dead center on the crankshaft), and the highest point in its travel, (top dead center on the crankshaft). This volume is generally expressed in cubic inches (in the United States), or in cubic centimeters (in England and Europe).

FORMULA: Piston Displacement = Area of piston X length of Stroke X No. of Cylinder.

Piston Displacement = Diameter 2 X .7854 X Stroke X No. of Cylinders.

NOTE: See Section 2 Fundamental problems on how the area of a piston is found.

EXAMPLE: Bore--------3.187
Stroke------3.75
Cylinders---8

Piston Displacement = 3.187^2 X .7854 X 3.75 X 8.

Piston Displacement = 239 cubic inches.

COMPRESSION RATIO

In order to find the compression ratio of an engine run one of the pistons to top dead center, then fill the clearance between the piston and the cylinder head with oil from a graduate or measure. It will be necessary to pour very carefully in order to get accurate measurement. This will give you the clearance volume in centimeters (most graduates measured by centimeters). It is then necessary to find the piston displacement, or swept volume, (if this has not already been done) in the same terms, (cubic centimeters), or convert the clearance to cubic inches. The formula used in finding compression ratio is:

$$\frac{V + C}{C}$$

EXAMPLE: V--Swept Volume
C--Clearance Volume
R--Compression Volumetric Ratio
A--Area of piston
16.39 cubic centimeter-1 cubic inch

Fill with oil and measure.
Bore 5 ins.

Stroke 8 ins.

Clearance measured by oil 245.85 Cubic Centimeters or 15 cubic inches.

Compression Ratio equals $Bore^2$ X .7854 X Stroke plus Clearance divided by Clearance Volume.

Compression Ratio equals (5^2 X .7854 X 8) plus 15 cubic inches divided by 15 cubic inches.

Compression Ratio   11.46:1

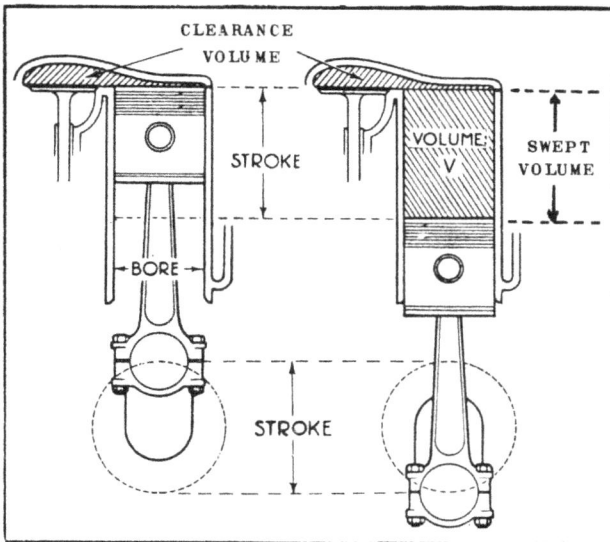

FIG. NO. 28 ABOVE: THIS DIAGRAM INDICATES THE APPLICATION OF BORE, STROKE, CLEARANCE VOLUME AND SWEPT VOLUME.
FIG. NO. 29 RIGHT: A SIMPLE TYPE OF CARBURETOR SHOWING THE RELATION OF FLOAT CHAMBER, AIR INTAKE, JET AND THROTTLE. MODERN CARBURETORS ARE SOMEWHAT MORE COMPLICATED, BUT FOLLOW THE SAME BASIC PRINCIPLE.

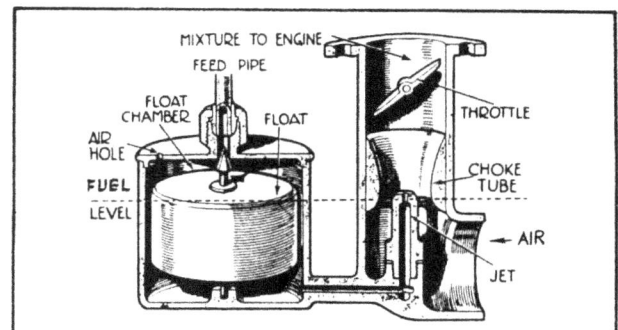

FUNDAMENTAL PROBLEMS

Following are some fundamental methods of figuring various problems which might arise while building a car, such as finding the volume of a tank, etc. It is hoped that these will be helpful to the builder. They will enable him to ascertain the amount of material needed for the job, and will help in similar building problems.

1. Circumference of a circle = Diameter X 3.1416.

Circumference equals 15 X 3.1416 equals 47.12 ins.

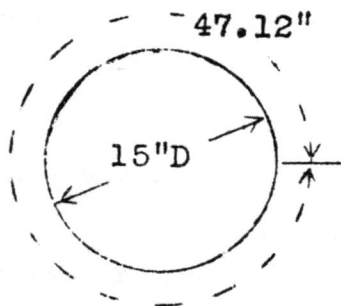

2. Area of a circle equals .7854 X diameter^2.

Area equals .7854 X 7^2 equals .7854 X 49 equals 38.48.

The formula is often expressed as 1/4 Pi D^2, or as radius squared X Pi (Pi equals 3.1416).

The above form Pi /4 equals .7854 is usually more convenient.

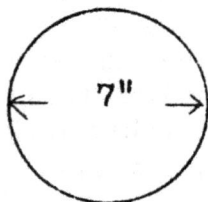

3. Area of a rectangle equals product of the two sides.

Area equals 5 ft. X 19 ft. equals 50 sq. ft.

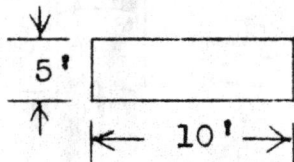

4. Closed rectangular tank: -- Find area of the six rectangles forming the sides, bottom, and top.

ENDS: Two rectangles 3 X 8 or 2 X 24 equals 48 sq. ft.

SIDES: Two rectangles 3 X 12 or 2 X 36 equals 72 sq. ft.

TOP & BOTTOM: Two rectangles 8 X 12 or 2 X 96 equals 192 sq. ft.

Total area of sheet metal required equals 312 sq. ft.

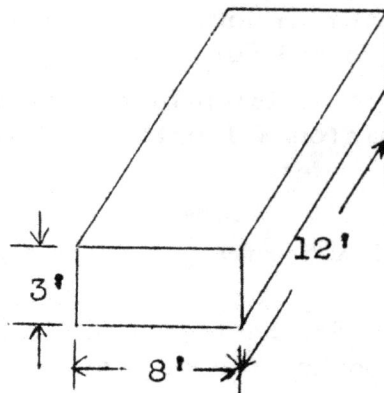

5. Cylindrical tank--area of two ends found as per (2) plus area of cylindrical part which is a rectangle if split along one side and flattened out.

EXAMPLE: Cylinder 6 in. diam., 10 ins. high, area of each end is: .7854 X 6 X 6 equals 28.3 sq. in.

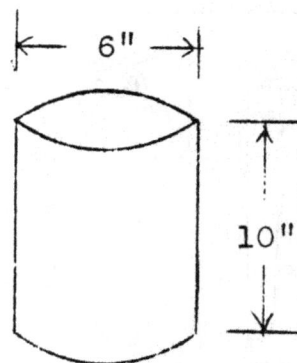

If cylindrical part were split and spread out flat, it would form a rectangle 10 ins. on one side, and the other side equal to the circumference.

(ILLUSTRATION FOR THIS EXAMPLE WILL BE FOUND ON NEXT PAGE.)

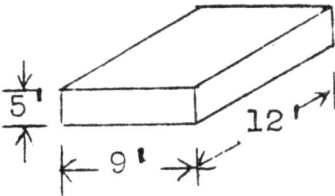

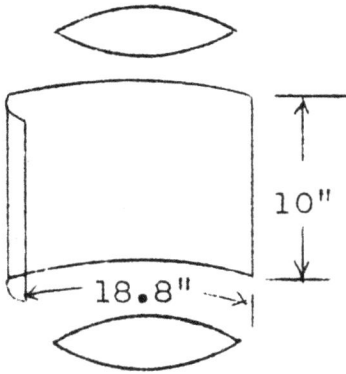

Circumference equals:
  6 X 3.1416 equals   18.84 sq. in.
Rectangular area equals:
  10 X 18.84 equals   188   sq. in.
Two ends equal:
  2 X 28.3 equals   57   sq. in.
Total area equals   245   sq. in.

6. Volume of rectangular tank equals length times width, times height.

Volume equals 9 X 5 X 12 equals 540 cu. ft.

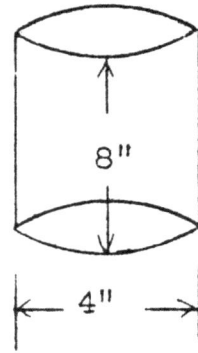

7. Volume of any tank in gallons found by multiplying cubic feet by 7.48. (Tank in (6) contains 540 X 7.48 equals 4039.20 gal.)

8. Volume of a cylinder equals area of base X height.

Area of base equals .7854 X 4 X 4 equals 12.56 sq. in.

Volume equals 12.56 X 8 equals 100.48 cu. in.

(This is necessary in finding piston displacement, capacity of tanks, etc.)

NOTE: 1 U. S. Gal. equals 231 cu. in.

TORQUE

Torque, or twisting effort, is often considered the rotary equivalent of force. It is expressed in terms of 'foot-pounds'. (The product of a force, times the lever action through which it is exerted)

EXAMPLE: In tightening a screwed pipe fitting, a plumber exerts a 50-pound pull on the end of a pipe wrench with a 4-foot handle. What torque is used in twisting the pipe fitting?

Torque equals 4 feet X 50 lbs. equals 200 foot-pounds.

If either the force or length of lever arm (wrench handle) were doubled, the torque would be doubled. The identical terms sometimes cause confusion between the foot-pound torque and the 'pound-foot' for torque, although the practice is not standardized enough to justify its use here.

An engine exerts a torque through its drive shaft. In order to simplify this idea of power transmission, suppose it is doing all its work by means of a dog protruding from the face of the flywheel, and engaging in a notch on the flange of some driven shaft, such as that of a dynamometer (a machine for measuring torque).

If the maximum torque the engine can exert is, for instance, 200 ft.-lbs. and the dog is located one foot out from the crankshaft center, it will obviously be transmitting a force of 200 lbs.

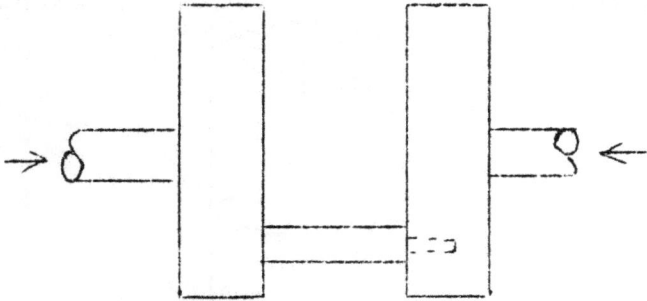

200 ft.-lbs. torque divided by 1 ft. radius equals 200 lbs. force.

It is possible to figure the brake horsepower of the engine from the torque and rpm, on the basis of the distance the above dog travels in one minute on its circular path.

Suppose the above engine is operating at 900 rpm when exerting its 200 ft.-lbs. torque. Then each minute a 200-lb. force is moved around in a circle having a 1-ft. radius (or 2-ft. diameter) 900 times.

Distance each revolution equals 2 X 3.14 equals 6.28 ft.

Distance in one minute equals 6.28 x 900 equals 5650 ft.

Work equals 5650 ft. x 200 lbs. equals 1,130,000 ft.-lbs. per min.

Power equals 1,130,000 divided by 33,000 equals 34.3 bhp.

STANDARD SPECIFICATIONS OF FORD ENGINES

V8-'60' - 1940 MODEL

Bore	2.6 ins.
Stroke	3.2 ins.
Piston Displacement	136 cubic ins.
Brake Horsepower	60 at 3500 RPM
Taxable Horsepower	21.6
Torque	94 ft. lbs. at 2500 RPM
Compression Ratio	6.6 : 1
Compression Pressure	158 lbs. at 2800 RPM
Compression Pres. at crg. speed	105 lbs.
Firing order	1-5-4-8-6-3-7-2

VALVE TIMING
(for 1940 V8 '60')

Intake opens	9.5° B.T.C.
Intake closes	54.5° A.B.C.
Exhaust opens	57.5° B.B.C.
Exhaust closes	6.5° A.T.C.
Valve clearance	.0125 - .0135

(This applies for all Ford V8 engines)

V8-'85' - 1940 MODEL

Bore	3.062 ins.
Stroke	3.75 ins.
Piston Displacement	221 cubic ins.
Brake Horsepower	85 at 3800 RPM
Taxable Horsepower	30
Torque	155 ft. lbs. at 2200 RPM
Compression Ratio	6.15 : 1
Compression Pressure	140 lbs. at 2400 RPM
Compression Pressure at crkg. speed	95 lbs.

VALVE TIMING
(for 1940 V8 '85')

Intake opens _____9.5° B.T.C.
Intake closes _____54.5° A.B.C.
Exhaust opens _____57.5° B.B.C.
Exhaust closes _____6.5° A.T.C.

V8-'100' - 1949 MODEL

Bore _____3.187 ins.
Stroke _____3.75 ins.
Piston Displacement _____239.4 cubic ins.
Brake Horsepower _____100 at 3,600 RPM
Taxable Horsepower _____32.5
Compression Ratio _____6.8 : 1

V8-'95' (Mercury) 1940 MODEL

Bore _____3.187 ins.
Stroke _____3.75 ins.
Piston Displacement _____239 cubic inches
Brake Horsepower _____95 at 3600 RPM
Taxable Horsepower _____32.5
Torque _____170 ft. lbs. 2100 RPM
Compression Ratio _____6.15 : 1
Compression Pressure _____145 lbs. at 2200 RPM
Compression Pressure at crkg. speed ____100 lbs.
Firing order _____1-5-4-8-6-3-7-2
                                            (This applies for
                                            all Ford V8 engines)

VALVE TIMING
(for 1940 Mercury)

Intake opens _____ 0.0° B.T.C.
Intake closes _____ 44.0° A.B.C.
Exhaust opens _____ 48.0° B.B.C.
Exhaust closes _____ 6.0° A.T.C.

V8-'110' (Mercury) 1949 MODEL

Bore _____ 3 3/16 ins.
Stroke _____ 4 ins.
Piston Displacement _____ 255.4 cubic ins.
Brake Horsepower _____ 110 at 3,600 RPM
Taxable Horsepower _____ 32.5
Maximum Torque _____ 200 ft. lbs. at 2,000 RPM
Compression Ratio _____ 6.8 : 1
Compression pressure _____ 170 lbs.
Compression pressure at crkg. speed_115 lbs.

VALVE TIMING
(for 1949 Mercury)

Intake opens _____10° B.T.C.
Intake closes _____50° A.B.C.
Exhaust opens _____50° B.B.C.
Exhaust closes _____10° A.T.C.

STOCK MODEL 'A' FORD ENGINE VALVE TIMING

Intake opens_____7.5° before T.D.C.
Intake closes_____48.5° after B.D.C.
Exhaust opens_____51.5 before B.D.C.
Exhaust closes_____4.5° after T.D.C.

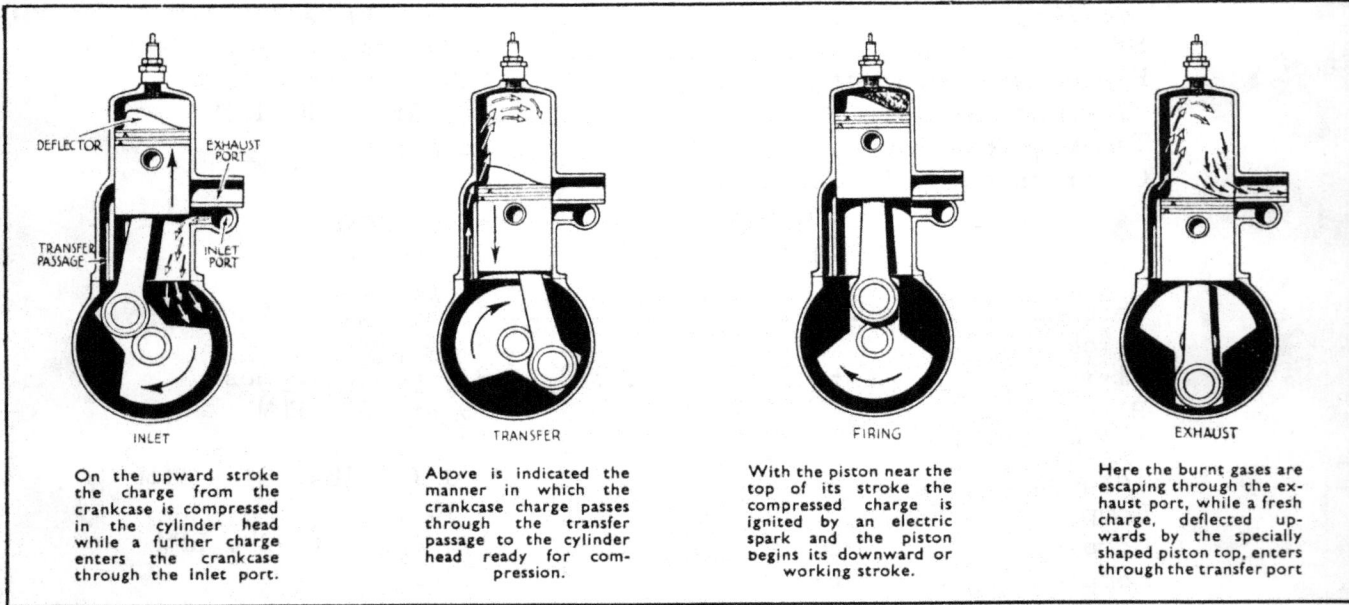

INLET

On the upward stroke the charge from the crankcase is compressed in the cylinder head while a further charge enters the crankcase through the inlet port.

TRANSFER

Above is indicated the manner in which the crankcase charge passes through the transfer passage to the cylinder head ready for compression.

FIRING

With the piston near the top of its stroke the compressed charge is ignited by an electric spark and the piston begins its downward or working stroke.

EXHAUST

Here the burnt gases are escaping through the exhaust port, while a fresh charge, deflected upwards by the specially shaped piston top, enters through the transfer port

FIG. NO. 32. THE OPERATION OF A TWO-STROKE ENGINE.

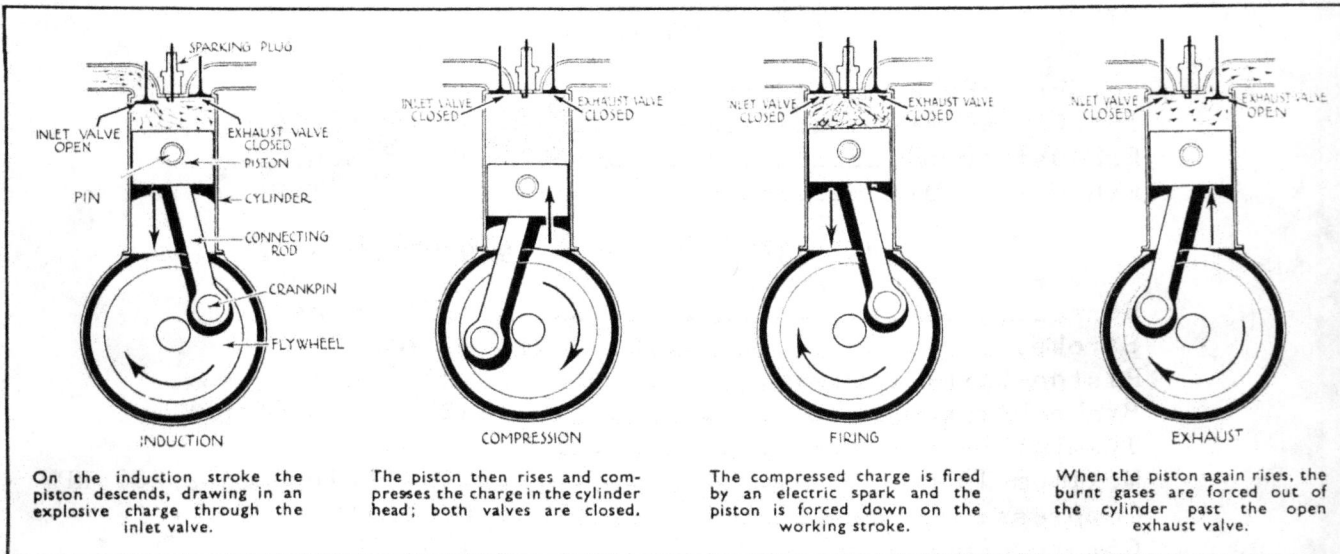

INDUCTION

On the induction stroke the piston descends, drawing in an **explosive** charge through the inlet valve.

COMPRESSION

The piston then rises and compresses the charge in the cylinder head; both valves are closed.

FIRING

The compressed charge is fired by an electric spark and the piston is forced down on the working stroke.

EXHAUST

When the piston again rises, the burnt gases are forced out of the cylinder past the open exhaust valve.

FIG. NO. 33.   THE OPERATION OF A FOUR-STROKE ENGINE.

MEASURING DEGREES ON UNMARKED FLYWHEELS

To convert degrees to distance on a flywheel of any diameter, multiply the diameter in inches by .1416. Then divide the circumference by 360.

For example, to lay out timing marks on a 14-inch flywheel, multiply 14 inches by 3.1416 to find the circumference (43.9824 inches). Dividing 43.9824 inches by 360 gives .122 inches for one degree.

If timing called for spark 30° before D.C., multiply 9.122 inches by 30, which is 3.66 inches. Then measure this distance (3.66 inches) before D.C. on the flywheel.

If the flywheel has starter teeth divide 360 by the number of teeth in the flywheel ring gear. This gives the value (in degrees) of each tooth. Thus on a 120 tooth flywheel the value of each tooth would be 3 degrees and the setting for 30° before D.C. would be 10 teeth before top dead center.

EXPLANATION OF ENGINE CAPACITY

In Europe or Great Britain the displacement of an engine is usually given in liters (or "litres"), whereas in the United States displacement is generally referred to in terms of cubic inches. Below is table for converting liters to cubic inches:

1/2 liter	30.5125 cu. in.	
1 "	61.025 " "	
1 1/2 "	91.5375 " "	
2 "	122.050 " "	
2 1/2 "	152.5625 " "	
3 "	183.073 " "	
3 1/2 "	213.5855 " "	
4 "	244.100 " "	
4 1/2 "	274.6125 " "	
5 "	305.125 " "	
5 1/2 "	330.6375 " "	
6 "	366.150 " "	
6 1/2 "	396.6625 " "	
7 "	427.175 " "	

(From the above table it will be noted that Duesenberg, Miller, and Offenhauser engines were built to the "liter" standards; i.e.,

91 cu. in.	1 1/2 liters
183 " "	3 "
274 " "	4 1/2 "

CUTTING LUBRICANTS

MATERIAL	DRILL	REAM	THREAD
Aluminum	Kerosene and Oil		
Babbitt	*Dry	Kerosene &	Dry
Brass	Dry	Turpentine	Dry
Cast Iron	Dry or Soda Water	Tallow & Graphite	Lard Oil
Copper	Lard Oil and Turpentine		
Monel	Red Lead and Oil		
Rawhide	Soap		
Steel	‡Soap or Soda Water		Lard Oil & Kerosene

*Kerosene and lard oil for boring.

‡Hard spots: 1 part powdered sulphur, 1 part oil, 2 parts castor oil.

HOW TO LOCATE EXACT POINT OF VALVE OPENING

Adjust the tappet clearance to 0.005 more than the specified timing clearance. Then turn the crankshaft in the direction of rotation until a 0.005 feeler is just gripped between the valve stem and lifter (or rocker arm).

For example, if the intake valve is supposed to open at top D.C. with 0.015 clearance, adjust the clearance to 0.020. Then insert a 0.005 feeler and turn the crankshaft until the feeler is just gripped. This would then be the same as the zero clearance position with the tappet set at 0.015

FORD V8 VALVE TIMINGS

(While reground cams vary somewhat with each supplier, common usage classifies the various types (approximately) in accordance with the following table for the Ford V8)

	LATE STOCK	EARLY STOCK	1949 MERC.	1/2 RACE	3/4 RACE	FULL RACE	SUPER (WINFIELD)	SUPER (HARMAN-COLLINS)	SUPER 'H' (HARMAN-COLLINS)
INTAKE OPENS (Deg. before T.D.C.)	0°	9½°	10°	21°	23°	26°	24°	28°	30°
INTAKE CLOSES (Deg. after B.D.C.)	44°	54½°	50°	59°	62°	64°	68°	67°	78°
EXHAUST OPENS (Deg. before B.D.C.)	48°	57½°	50°	54°	56°	59°	68°	61°	64°
EXHAUST CLOSES (Deg. after T.D.C.)	6°	6½°	10°	16°	19°	21°	24°	24°	26°
INTAKE DURATION (Deg. crankshaft)	224°	244°	240°	260°	265°	270°	272°	275°	288°
EXHAUST DURATION (Deg. crankshaft)	234°	244°	240°	250°	255°	260°	272°	265°	270°
OVERLAP (Deg. intake and exhaust are open at the same time)	6°	16°	20°	37°	42°	47°	48°	52°	56°

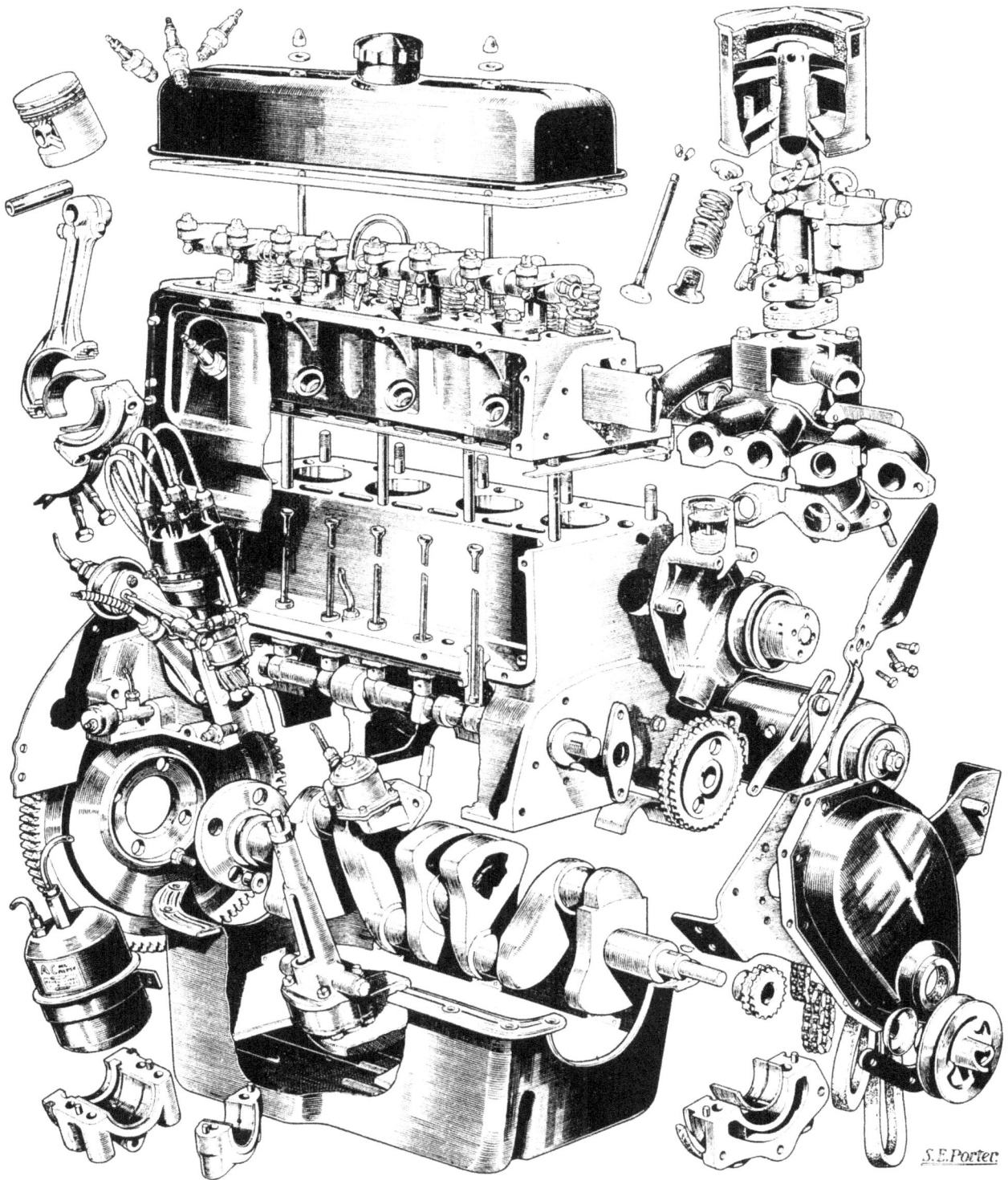

S.E.Porter

FIG. NO. 34. ANOTHER BRITISH ILLUSTRATION SHOWING AN "EXPLODED" VIEW OF A 4-CYLINDER OVERHEAD VALVE (PUSH-ROD) ENGINE. IT IS SHOWN HERE BECAUSE IT ILLUSTRATES ALL OPERATING PARTS WITH EXCEPTIONAL CLARITY.

CONVERSION TABLE

MILLIMETRES TO INCHES

mm. Ins.									
	15 = .5905	30 = 1.1811	45 = 1.7716	60 = 2.3622	75 = 2.9527	90 = 3.5433	105 = 4.1338	120 = 4.7244	135 = 5.3149
.25 = .0098	15.25 = .6004	30.25 = 1.1909	45.25 = 1.7815	60.25 = 2.3720	75.25 = 2.9626	90.25 = 3.5531	105.25 = 4.1437	120.25 = 4.7342	135.25 = 5.3248
.50 = .0197	15.50 = .6102	30.50 = 1.2008	45.50 = 1.7913	60.50 = 2.3819	75.50 = 2.9724	90.50 = 3.5630	105.50 = 4.1535	120.50 = 4.7441	135.50 = 5.3346
.75 = .0295	15.75 = .6201	30.75 = 1.2106	45.75 = 1.8012	60.75 = 2.3917	75.75 = 2.9823	90.75 = 3.5728	105.75 = 4.1634	120.75 = 4.7539	135.75 = 5.3445
1 = .0394	16 = .6299	31 = 1.2205	46 = 1.8110	61 = 2.4016	76 = 2.9921	91 = 3.5827	106 = 4.1732	121 = 4.7638	136 = 5.3543
1.25 = .0492	16.25 = .6398	31.25 = 1.2303	46.25 = 1.8209	61.25 = 2.4114	76.25 = 3.0020	91.25 = 3.5925	106.25 = 4.1831	121.25 = 4.7736	136.25 = 5.3642
1.50 = .0591	16.50 = .6496	31.50 = 1.2402	46.50 = 1.8307	61.50 = 2.4213	76.50 = 3.0118	91.50 = 3.6024	106.50 = 4.1929	121.50 = 4.7835	136.50 = 5.3740
1.75 = .0689	16.75 = .6594	31.75 = 1.2500	46.75 = 1.8405	61.75 = 2.4311	76.75 = 3.0216	91.75 = 3.6122	106.75 = 4.2027	121.75 = 4.7933	136.75 = 5.3838
2 = .0787	17 = .6693	32 = 1.2598	47 = 1.8504	62 = 2.4409	77 = 3.0315	92 = 3.6220	107 = 4.2126	122 = 4.8031	137 = 5.3937
2.25 = .0886	17.25 = .6791	32.25 = 1.2697	47.25 = 1.8602	62.25 = 2.4508	77.25 = 3.0413	92.25 = 3.6319	107.25 = 4.2224	122.25 = 4.8130	137.25 = 5.4035
2.50 = .0984	17.50 = .6890	32.50 = 1.2795	47.50 = 1.8701	62.50 = 2.4606	77.50 = 3.0512	92.50 = 3.6417	107.50 = 4.2323	122.50 = 4.8228	137.50 = 5.4134
2.75 = .1083	17.75 = .6988	32.75 = 1.2894	47.75 = 1.8799	62.75 = 2.4705	77.75 = 3.0610	92.75 = 3.6516	107.75 = 4.2421	122.75 = 4.8327	137.75 = 5.4232
3 = .1181	18 = .7087	33 = 1.2992	48 = 1.8898	63 = 2.4803	78 = 3.0709	93 = 3.6614	108 = 4.2520	123 = 4.8425	138 = 5.4331
3.25 = .1280	18.25 = .7185	33.25 = 1.3091	48.25 = 1.8996	63.25 = 2.4901	78.25 = 3.0807	93.25 = 3.6713	108.25 = 4.2618	123.25 = 4.8524	138.25 = 5.4429
3.50 = .1378	18.50 = .7283	33.50 = 1.3189	48.50 = 1.9094	63.50 = 2.5000	78.50 = 3.0905	93.50 = 3.6811	108.50 = 4.2716	123.50 = 4.8622	138.50 = 5.4527
3.75 = .1476	18.75 = .7382	33.75 = 1.3287	48.75 = 1.9193	63.75 = 2.5098	78.75 = 3.1004	93.75 = 3.6909	108.75 = 4.2815	123.75 = 4.8720	138.75 = 5.4625
4 = .1575	19 = .7480	34 = 1.3386	49 = 1.9291	64 = 2.5197	79 = 3.1102	94 = 3.7008	109 = 4.2913	124 = 4.8819	139 = 5.4724
4.25 = .1673	19.25 = .7579	34.25 = 1.3484	49.25 = 1.9390	64.25 = 2.5295	79.25 = 3.1201	94.25 = 3.7106	109.25 = 4.3012	124.25 = 4.8917	139.25 = 5.4823
4.50 = .1772	19.50 = .7677	34.50 = 1.3583	49.50 = 1.9488	64.50 = 2.5394	79.50 = 3.1299	94.50 = 3.7205	109.50 = 4.3110	124.50 = 4.9016	139.50 = 5.4921
4.75 = .1870	19.75 = .7776	34.75 = 1.3681	49.75 = 1.9587	64.75 = 2.5492	79.75 = 3.1398	94.75 = 3.7303	109.75 = 4.3209	124.75 = 4.9114	139.75 = 5.5020
5 = .1968	20 = .7874	35 = 1.3779	50 = 1.9685	65 = 2.5590	80 = 3.1496	95 = 3.7401	110 = 4.3307	125 = 4.9212	140 = 5.5118
5.25 = .2067	20.25 = .7972	35.25 = 1.3878	50.25 = 1.9783	65.25 = 2.5689	80.25 = 3.1594	95.25 = 3.7500	110.25 = 4.3405	125.25 = 4.9311	140.25 = 5.5216
5.50 = .2165	20.50 = .8071	35.50 = 1.3976	50.50 = 1.9882	65.50 = 2.5787	80.50 = 3.1693	95.50 = 3.7598	110.50 = 4.3504	125.50 = 4.9409	140.50 = 5.5315
5.75 = .2264	20.75 = .8169	35.75 = 1.4075	50.75 = 1.9980	65.75 = 2.5886	80.75 = 3.1791	95.75 = 3.7697	110.75 = 4.3602	125.75 = 4.9508	140.75 = 5.5413
6 = .2362	21 = .8268	36 = 1.4173	51 = 2.0079	66 = 2.5984	81 = 3.1890	96 = 3.7795	111 = 4.3701	126 = 4.9606	141 = 5.5512
6.25 = .2461	21.25 = .8366	36.25 = 1.4272	51.25 = 2.0177	66.25 = 2.6083	81.25 = 3.1988	96.25 = 3.7894	111.25 = 4.3799	126.25 = 4.9705	141.25 = 5.5610
6.50 = .2559	21.50 = .8465	36.50 = 1.4370	51.50 = 2.0276	66.50 = 2.6181	81.50 = 3.2087	96.50 = 3.7992	111.50 = 4.3898	126.50 = 4.9803	141.50 = 5.5709
6.75 = .2657	21.75 = .8563	36.75 = 1.4468	51.75 = 2.0374	66.75 = 2.6279	81.75 = 3.2185	96.75 = 3.8090	111.75 = 4.3996	126.75 = 4.9901	141.75 = 5.5807
7 = .2756	22 = .8661	37 = 1.4567	52 = 2.0472	67 = 2.6378	82 = 3.2283	97 = 3.8189	112 = 4.4094	127 = 5.0000	142 = 5.5905
7.25 = .2854	22.25 = .8760	37.25 = 1.4665	52.25 = 2.0571	67.25 = 2.6476	82.25 = 3.2382	97.25 = 3.8287	112.25 = 4.4193	127.25 = 5.0098	142.25 = 5.6004
7.50 = .2953	22.50 = .8858	37.50 = 1.4764	52.50 = 2.0669	67.50 = 2.6575	82.50 = 3.2480	97.50 = 3.8386	112.50 = 4.4291	127.50 = 5.0197	142.50 = 5.6102
7.75 = .3051	22.75 = .8957	37.75 = 1.4862	52.75 = 2.0768	67.75 = 2.6673	82.75 = 3.2579	97.75 = 3.8484	112.75 = 4.4390	127.75 = 5.0295	142.75 = 5.6201
8 = .3150	23 = .9055	38 = 1.4961	53 = 2.0866	68 = 2.6772	83 = 3.2677	98 = 3.8583	113 = 4.4488	128 = 5.0394	143 = 5.6299
8.25 = .3248	23.25 = .9153	38.25 = 1.5059	53.25 = 2.0965	68.25 = 2.6870	83.25 = 3.2776	98.25 = 3.8681	113.25 = 4.4587	128.25 = 5.0492	143.25 = 5.6398
8.50 = .3346	23.50 = .9252	38.50 = 1.5157	53.50 = 2.1063	68.50 = 2.6968	83.50 = 3.2874	98.50 = 3.8779	113.50 = 4.4685	128.50 = 5.0590	143.50 = 5.6496
8.75 = .3445	23.75 = .9350	38.75 = 1.5256	53.75 = 2.1161	68.75 = 2.7067	83.75 = 3.2972	98.75 = 3.8878	113.75 = 4.4783	128.75 = 5.0689	143.75 = 5.6594
9 = .3543	24 = .9449	39 = 1.5354	54 = 2.1260	69 = 2.7165	84 = 3.3071	99 = 3.8976	114 = 4.4882	129 = 5.0787	144 = 5.6693
9.25 = .3642	24.25 = .9547	39.25 = 1.5453	54.25 = 2.1358	69.25 = 2.7264	84.25 = 3.3169	99.25 = 3.9075	114.25 = 4.4980	129.25 = 5.0886	144.25 = 5.6791
9.50 = .3740	24.50 = .9646	39.50 = 1.5551	54.50 = 2.1457	69.50 = 2.7362	84.50 = 3.3268	99.50 = 3.9173	114.50 = 4.5079	129.50 = 5.0984	144.50 = 5.6890
9.75 = .3839	24.75 = .9744	39.75 = 1.5650	54.75 = 2.1555	69.75 = 2.7461	84.75 = 3.3366	99.75 = 3.9272	114.75 = 4.5177	129.75 = 5.1083	144.75 = 5.6989
10 = .3937	25 = .9842	40 = 1.5748	55 = 2.1653	70 = 2.7559	85 = 3.3464	100 = 3.9370	115 = 4.5275	130 = 5.1181	145 = 5.7086
10.25 = .4035	25.25 = .9941	40.25 = 1.5846	55.25 = 2.1752	70.25 = 2.7657	85.25 = 3.3563	100.25 = 3.9468	115.25 = 4.5374	130.25 = 5.1279	145.25 = 5.7185
10.50 = .4134	25.50 = 1.0039	40.50 = 1.5945	55.50 = 2.1850	70.50 = 2.7756	85.50 = 3.3661	100.50 = 3.9567	115.50 = 4.5472	130.50 = 5.1378	145.50 = 5.7283
10.75 = .4232	25.75 = 1.0138	40.75 = 1.6043	55.75 = 2.1949	70.75 = 2.7854	85.75 = 3.3760	100.75 = 3.9665	115.75 = 4.5571	130.75 = 5.1476	145.75 = 5.7382
11 = .4331	26 = 1.0236	41 = 1.6142	56 = 2.2047	71 = 2.7953	86 = 3.3858	101 = 3.9764	116 = 4.5669	131 = 5.1575	146 = 5.7480
11.25 = .4429	26.25 = 1.0335	41.25 = 1.6240	56.25 = 2.2146	71.25 = 2.8051	86.25 = 3.3957	101.25 = 3.9862	116.25 = 4.5768	131.25 = 5.1673	146.25 = 5.7579
11.50 = .4528	26.50 = 1.0433	41.50 = 1.6339	56.50 = 2.2244	71.50 = 2.8150	86.50 = 3.4055	101.50 = 3.9961	116.50 = 4.5866	131.50 = 5.1772	146.50 = 5.7677
11.75 = .4626	26.75 = 1.0531	41.75 = 1.6437	56.75 = 2.2342	71.75 = 2.8248	86.75 = 3.4153	101.75 = 4.0059	116.75 = 4.5964	131.75 = 5.1870	146.75 = 5.7775
12 = .4724	27 = 1.0630	42 = 1.6535	57 = 2.2441	72 = 2.8346	87 = 3.4252	102 = 4.0157	117 = 4.6063	132 = 5.1968	147 = 5.7874
12.25 = .4823	27.25 = 1.0728	42.25 = 1.6634	57.25 = 2.2539	72.25 = 2.8445	87.25 = 3.4350	102.25 = 4.0256	117.25 = 4.6161	132.25 = 5.2067	147.25 = 5.7972
12.50 = .4921	27.50 = 1.0827	42.50 = 1.6732	57.50 = 2.2638	72.50 = 2.8543	87.50 = 3.4449	102.50 = 4.0354	117.50 = 4.6260	132.50 = 5.2165	147.50 = 5.8071
12.75 = .5020	27.75 = 1.0925	42.75 = 1.6831	57.75 = 2.2736	72.75 = 2.8642	87.75 = 3.4547	102.75 = 4.0453	117.75 = 4.6358	132.75 = 5.2264	147.75 = 5.8169
13 = .5118	28 = 1.1024	43 = 1.6929	58 = 2.2835	73 = 2.8740	88 = 3.4646	103 = 4.0551	118 = 4.6457	133 = 5.2362	148 = 5.8268
13.25 = .5217	28.25 = 1.1122	43.25 = 1.7028	58.25 = 2.2933	73.25 = 2.8839	88.25 = 3.4744	103.25 = 4.0650	118.25 = 4.6555	133.25 = 5.2461	148.25 = 5.8366
13.50 = .5315	28.50 = 1.1220	43.50 = 1.7126	58.50 = 2.3031	73.50 = 2.8937	88.50 = 3.4842	103.50 = 4.0748	118.50 = 4.6653	133.50 = 5.2559	148.50 = 5.8464
13.75 = .5413	28.75 = 1.1319	43.75 = 1.7224	58.75 = 2.3130	73.75 = 2.9035	88.75 = 3.4941	103.75 = 4.0846	118.75 = 4.6752	133.75 = 5.2657	148.75 = 5.8563
14 = .5512	29 = 1.1417	44 = 1.7323	59 = 2.3228	74 = 2.9134	89 = 3.5039	104 = 4.0945	119 = 4.6850	134 = 5.2756	149 = 5.8661
14.25 = .5610	29.25 = 1.1516	44.25 = 1.7421	59.25 = 2.3327	74.25 = 2.9232	89.25 = 3.5138	104.25 = 4.1043	119.25 = 4.6949	134.25 = 5.2854	149.25 = 5.8760
14.50 = .5709	29.50 = 1.1614	44.50 = 1.7520	59.50 = 2.3425	74.50 = 2.9331	89.50 = 3.5236	104.50 = 4.1142	119.50 = 4.7047	134.50 = 5.2953	149.50 = 5.8858
14.75 = .5807	29.75 = 1.1713	44.75 = 1.7618	59.75 = 2.3524	74.75 = 2.9429	89.75 = 3.5335	104.75 = 4.1240	119.75 = 4.7146	134.75 = 5.3051	149.75 = 5.8957

APPROXIMATE WEIGHT OF VARIOUS LIQUIDS

Liquid	Pounds Per Gal.
Alcohol	6.576
Gasoline	6.042
Kerosene	6.668
Oil (lubricating)	7.584
Water	8.335

ELECTRICAL CALCULATIONS

To find:

Amperes, divide volts by ohms.

Volts, multiply amperes by ohms.

Ohms, divide volts by amperes.

Watts, multiply volts by amperes (one kilowatt is 1000 watts).

Horsepower, divide watts by 746 or kilowatts by 0.746. One kilowatt equals 1.340 horsepower.

EFFICIENCY

ECONOMY

POWER

PERFORMANCE

THE PERFECT COMBINATION

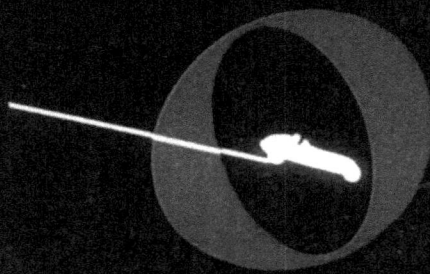

Offenhauser EQUIPMENT

CAST ALUMINUM HIGH COMPRESSION
FORD & MERCURY CYLINDER HEADS
1939 TO 1953

MANUFACTURED BY

Offenhauser

In the selection of a power head, do not be mislead by high compression ratios. It is true, that if every power factor involved were 100% efficient, then the higher the compression ratio, the more power developed.

However, where the engine is not custom made, it very often proves that a lower compression ratio will produce a higher overall efficiency.

- EXTRA HEAVY ALUMINUM ALLOY CONSTRUCTION.
- PRECISION CAST WITH MAXIMUM RIB DESIGN.
- COOLER RUNNING WITH MORE WATER CAPACITY.
- MORE FIN AREA FOR RAPID HEAT DISSIPATION.
- WATER JACKETS CLOSE TO CRITICAL HEAT POINTS.
- FINEST AVAILABLE FOR STREET, HIGHWAY OR COMPETITION USE.
- EXTREMELY POPULAR BECAUSE OF SUPERIOR WORKMANSHIP AND PROVEN QUALITY.

TO ORDER HEADS
PART NO.
DESIGNATES
P A I R

1949 – 1953

Catalog No. 1069
See Chart - Specify Ratio

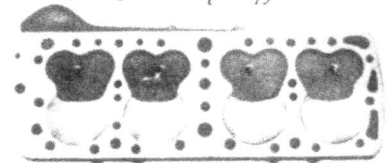

1939 – 1948

Catalog No. 1068
See Chart - Specify Ratio

V8 – 60

Catalog No. 1070
See Chart - Specify Ratio

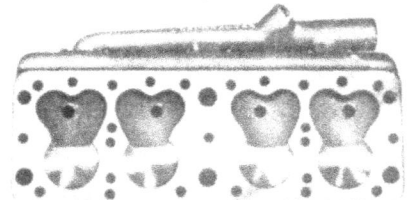

COMPRESSION RATIO CHART

| Engine Size | | | | Head Numbers | | | Cubic Inch |
Bore	Stroke	No. 425	No. 400	No. 375	No. 350	No. 325	Displacement
3-1/16	3-3/4	7.1	7.6	7.9	8.5	9.2	220.92
3-1/16	3-7/8	7.2	7.7	8.2	8.8	9.5	228.28
3-1/16	4	7.4	7.9	8.4	9.	9.8	235.648
3-3/16	3-3/4	7.4	7.9	8.5	9.2	9.9	239.312
3-3/16	3-7/8	7.7	8.2	8.8	9.4	10.2	247.288
3-3/16	4	8.0	8.5	9.	9.7	10.5	255.272
3-3/16	4-1/8	8.2	8.7	9.3	9.9	10.8	263.24
3-5/16	3-3/4	8.1	8.6	9.1	9.8	10.6	258.48
3-5/16	3-7/8	8.3	8.8	9.4	10.1	10.9	267.096
3-5/16	4	8.6	9.1	9.7	10.4	11.3	275.712
3-5/16	4-1/8	8.8	9.3	9.9	10.7	11.6	284.328
3-3/8	3-3/4	8.3	8.8	9.4	10.1	10.9	268.376
3-3/8	3-7/8	8.6	9.1	9.7	10.4	11.3	277.328
3-3/8	4	8.9	9.4	10.	10.7	11.6	286.272
3-3/8	4-1/8	9.1	9.6	10.3	11.1	11.9	295.20

The above approximate compression ratios are figured on non-relieved blocks, ratio is lowered depending upon depth of relief.

EXAMPLE: No. 400 indicates .400 valve clearance

| V8-60 Engine | | Head Numbers | | | Cubic Inch |
Bore	Stroke	No. 300	No. 275	No. 255	Displacement
2.600	3.200	9.5	10.5	11.5	135.912

FOR SPECIAL COMPRESSION
RATIOS, ADVISE BORE, STROKE,
AND WHETHER BLOCK IS RELIEVED.

ALL PRICES ON ATTACHED LIST -- USE CATALOG NUMBERS WHEN ORDERING

FOR POLISHED EQUIPMENT -- SEE POLISHING PRICE LIST

1932 TO 1953
FORD AND MERCURY
INTAKE MANIFOLDS

MANUFACTURED BY

Offenhauser

Offenhauser intake manifolds are cast of high grade aluminum alloy, expertly engineered and precision machined . . . . manifolds provide increased power, economical operation, and a smoother, better balanced engine.

Equalized fuel distribution to all cylinders allows engine to operate at maximum efficiency throughout all stages of RPM . . . . in addition the custom appearance of our products is a great improvement where pride of ownership is prevalent.

1949 - 53 Regular dual manifold . . . . Catalog No. 1075
Outstanding feature of our most popular manifold is its simple installation procedure. Has almost same performance as our super dual manifolds. Generator fits in original position, with no special fan belts. Stock fuel pump is O.K. Necessary to use this manifold with 49 - 53 Ford carburetors or Stromberg No. 380272.
Linkage & Fuel Line Installation Kit
   1949-53 Ford . . . . Catalog No. 1086
   1949 Mercury . . . . Catalog No. 1086-A
   1950-53 Mercury . . . . Catalog No. 1086-B

1949 - 53 Super dual manifold . . . . Catalog No. 1076
   Complete with generator bracket.
   (Not recommended for use with automatic transmission.)
Universal Linkage & Fuel Line
   Ford & Mercury Installation Kit . . . . Catalog No. 1087

1949 - 53 Triple manifold . . . . Catalog No. 1077
   Complete with generator bracket.
   (Not recommended for use with automatic transmission.)
Universal Linkage & Fuel Line
   Ford & Mercury Installation Kit . . . . Catalog No. 1085

1949 - 53 Single carburetor four-throat manifold . . . . . . Catalog No. 1078
Designed to be used with Cadillac, Oldsmobile or 1953 Lincoln Four-Throat carburetors. *(Absolutely required to use a full centrifugal advance mechanism distributor, which has no vacuum control such as Mallory YC275A. Adequate air filter necessary.)*
Universal Linkage & Fuel Line
   Ford & Mercury Installation Kit . . . . Catalog No. 1089

1942 - 48 Super dual manifold . . . . Catalog No. 1073
   Complete with generator and fan carrier brackets.
Universal Linkage & Fuel Line
   Ford & Mercury Installation Kit . . . . Catalog No. 1084

1942 - 48 Triple manifold . . . . Catalog No. 1074
   Complete with generator and fan carrier brackets.
Universal Ford & Mercury
   Linkage & Fuel Line Installation Kit . . . . Catalog No. 1085

1932 - 48 Regular dual manifold . . . . Catalog No. 1090
This is one of our newest items, easily installed. Universal Ford & Mercury Fuel Line Linkage Kit . . . . Catalog No. 1084-A

1932 - 48 Single Carburetor Four-Throat manifold . . . . . Catalog No. 1079
Designed to adapt Cadillac, Oldsmobile or 1953 Lincoln four-throat carburetor to the Ford and Mercury. Use any type distributor for this one. Adequate air filter necessary.
Universal Ford & Mercury
   Linkage & Fuel Line Installation Kit . . . . Catalog No. 1088

1932 - 41 Super dual manifold . . . . Catalog No. 1071
   Complete with generator bracket.
Universal Ford & Mercury
   Fuel Line and Linkage Installation Kit . . . . Catalog No. 1084

1932 - 41 Triple manifold . . . . Catalog No. 1072
   Complete with generator bracket.
Universal Ford & Mercury
   Fuel Line & Linkage Installation Kit . . . . Catalog No. 1085

Offenhauser FUEL LINE AND LINKAGE KITS

1949-53 Reg. dual . . . Cat. No. 1086	1932-48 Super dual . . Cat. No. 1084
1949-53 Super dual . . Cat. No. 1087	1932-48 4-throat . . . . Cat. No. 1088
1949-53 4-throat . . . . Cat. No. 1089	1932-53 Triple . . . . . Cat. No. 1085

1949 - 53 . . . . . . Catalog No. 1075
1932 - 48 . . . . . . Catalog No. 1090

1949 - 53 . . . . . . Catalog No. 1076

1949 - 53 . . . . . . Catalog No. 1077

1949 - 53 . . . . . . Catalog No. 1078
1932 - 48 . . . . . . Catalog No. 1079

1942 - 48 . . . . . . Catalog No. 1073

1942 - 48 . . . . . . Catalog No. 1074

1932 - 41 . . . . . . Catalog No. 1071

1932 - 41 . . . . . . Catalog No. 1072

INTAKE MANIFOLDS
AND VALVE COVERS

Offenhauser intake manifolds are cast of high grade aluminum alloy, expertly engineered and precision machined . . . . manifolds provide increased power, economical operation, and a smoother, better balanced engine.

Equalized fuel distribution to all cylinders allows engine to operate at maximum efficiency throughout all stages of RPM . . . . in addition the custom appearance of our products is a great improvement where pride of ownership is prevalent.

All manifolds on this page are complete with gaskets, fuel lines, throttle linkages and instructions included at no extra charge.

Our valve covers have special finned-top, cast-aluminum alloy and highly polished for added engine beauty. Valve covers help reduce engine heat and rocker arm noise.

CHEVROLET 1937-54 (Now with oil filter mounting)

	Cat. No.
Dual manifold, Std. Trans. 1937-1952	1034
Dual manifold, Power Glide - 1950-1952	1035
Dual manifold, Std. & Power Glide - 1953-56	1035
Triple manifold, Std. Trans. 1937-1952	1177
Triple manifold, Power Glide - 1950-1952	1409
Triple manifold, Std. & Power Glide - 1953-56	1409
Valve Cover, polished - 1937-53	1036
Valve Cover, polished - 1954-56	2731

FORD 6 OHV 1952-56

Dual manifold, Std. Trans. - 1952-53	1082
Dual manifold, Automatic Trans. - 1952-53	2462
Dual manifold, Std. Trans. - 1954-56	2728
Dual manifold, Automatic Trans. - 1954-56	2729
Triple manifold, Std. Trans. - 1952-53	3126
Triple manifold, Automatic Trans. - 1952-53	3127
Triple manifold, Std. Trans. - 1954-56	3128
Triple manifold, Automatic Trans. - 1954-56	3129
Valve Cover, polished	1083

CADILLAC V8 1949-56

Dual manifold	1080
Valve Cover, polished per pair	1037

STUDEBAKER V8 1951-54

Manifold, adapts 4-throat carburetors	2262
Valve Covers, polished per pair	1190

STUDEBAKER CHAMPION 1939-56

Dual manifold - 1939-52	2706
Dual manifold - 1953-56	2708

DODGE V8 1953-56 except 500 Series

Dual manifold	1186
Quad. manifold, adapts 4-throat carburetors	2257
Valve Cover Caps, polished, per pair	1189

PLYMOUTH V8 1955 only

Dual manifold	3130

PLYMOUTH & DODGE 6

Dual manifold, Plymouth 1937-56	2691
Dual manifold, Dodge 6 1938-56	2374
Dual manifold, Dodge 6 1938-1954 (½ to 1½ Ton Truck	2375

V8 OHV FORD, MERCURY & THUNDERBIRD

Triple-Dual manifold, furnished carburetor pad cover easily removed to quickly convert to triple manifold.

Interchangeable linkage and fuel lines included	2726
1954 Ford and Mercury	2726
1955-56 Ford, Mercury and Thunderbird	3199
1957 Ford, Mecury, and Thunderbird	3199-7
Valve Covers, polished per pair 1954-57	2727

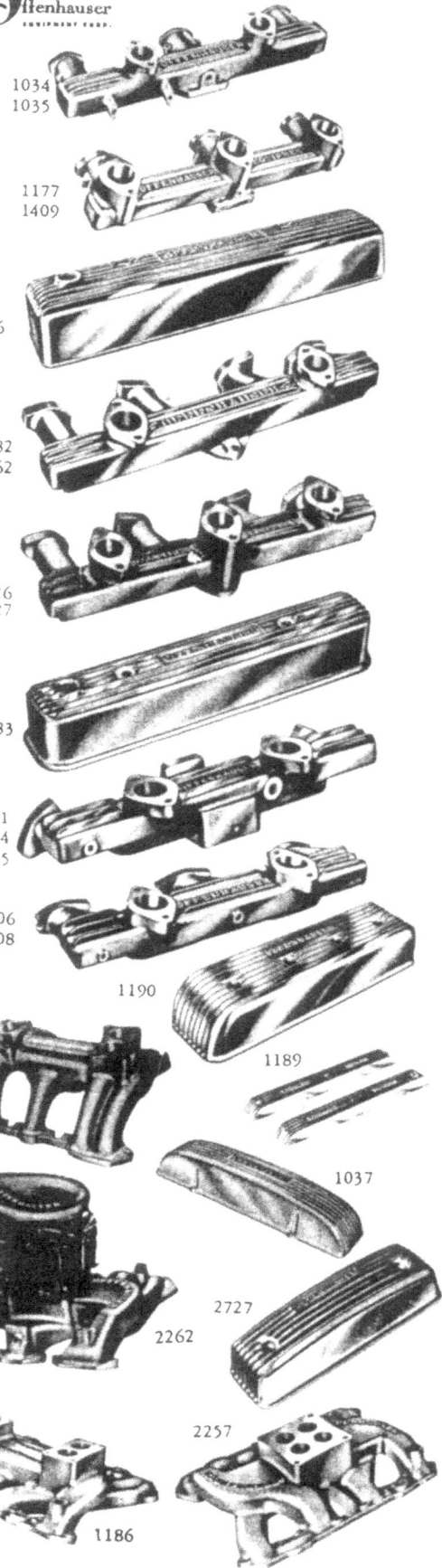

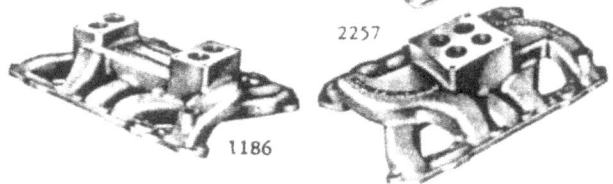

1034
1035

1177
1409

1036

1082
2462

3126
3127

1083

2691
2374
2375

2706
2708

1190

1080

1189

1037

2727

2262

2257

1186

2726
&
3199

For 1955 thru 1957 ...
315 cubic inch and 500 Engines

This manifold is now being offered as optional equipment on the 1957 Dodge D500 engines from the Dodge Main Plant in Detroit.

DUAL QUAD

Manifold only	#3614	$82.50
Installation Kit	#3616	$15.95

TRIPLE CARBURETOR

This unit may be ordered for either stock 4-bolt carburetors or Stromberg and Holley 3-bolt type. Be sure to specify.

Manifold only	#3615	$82.50
Installation Kit 4-bolt Carburetors	#3617	$15.95
Installation Kit 3-bolt Stromberg	#3618	$15.95
Installation Kit 3-bolt Holley	#3619	$21.00
Installation Kit 3-bolt Stromberg with Selective Linkage	#3620	$23.95
Installation Kit 3-bolt Holley with Selective Linkage	#3621	$25.25

Description	Cat. No.
Manifold only	3371
Standard throttle Linkage	1173
Ball bearing throttle Linkage	2866
Fuel block kit	2734

4

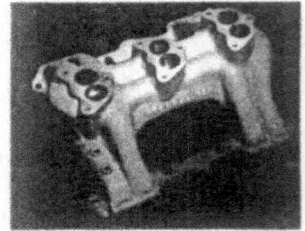

Cadillac 1949–56
No. 3371

DODGE 1953–56   PLYMOUTH 1955

(All except the 315 cubic inch Dodge engine)

No. 3372

Description	Cat. No.
Manifold only	3372
Ball bearing throttle Linkage	2866
Standard throttle Linkage	1173
Fuel block kit	2734

BUICK V8

Dual Quad — with installation kit.

1953–56	No. 3412
1957–58	No. 3556

Triple — when ordering specify whether for 4 Bolt or 3 Bolt Carburetors — less kit.

1953–56	No. 3413
1957–58	No. 3557

Manifold Kits for triple Buick

4 Bolt Kit	No. 3414
3 Bolt Kit	No. 3415

Buick Valve Covers

1953–58	No. 3416

SPORTS CAR OWNERS

BEAUTIFUL VALVE COVER
for TRIUMPH TR-2
also TRIUMPH RENOWNED, DORETTI, MORGAN PLUS 4 *and* STD. VANGUARD

The ribbed cover dresses up and tones down the powerful little engine. Uses stock gaskets, positive oil seal maintained through use of built-in gasket retainers. Comes complete with two chromed acorn hold down bolts and chromed oil breather cap.

Part No. 3132

OLDSMOBILE

Dual Quad Manifolds

1949–53	No. 3200
1954–56	No. 3285

Oldsmobile Valve Covers

1949–56	No. 3286
1957–58	No. 3286–7

CHEVROLET V8

Manifold	No. 3287
Valve Covers	No. 3288

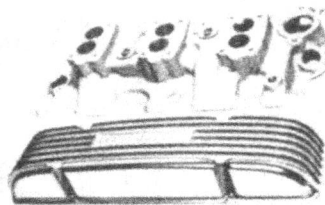

CHEVROLET VALVE COVERS

1955–58	No. 3288

CHEVROLET V8 COMBINATION DUAL TRIPLE MANIFOLD

Incorporating the famous and ORIGINAL Offenhauser dual-triple design. Winning two-way combination. Dual set-up for street use. Triple manifold for competition. Performance proven for high speed. Delivers all the power. Increases engine efficiency. Offenhauser original DUAL-TRIPLE MANIFOLD comes complete with kit and carburetor pad cover. Available for either 3 Bolt or 4 Bolt Carburetors. Specify when ordering.

1955–56	No. 3287	Fits Chevrolet V8 engines except
1957–58	No. 3558	1958 – 348 cu. inch

THE PERFECT COMBINATION

6 CARBURETOR MANIFOLDS

At last a line of Manifolds that have been engineered. These units incorporate the famous "Offenhauser Know How". Use with progressive-type linkage from 2 to 6 carburetors for street use, then all out for drag strip.

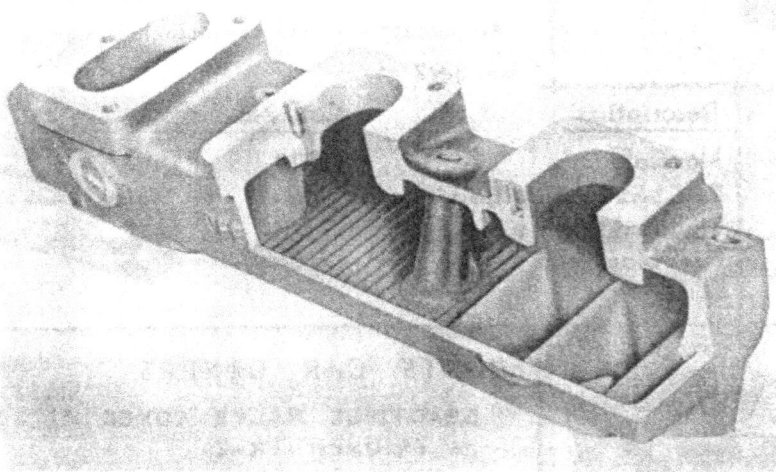

- BALANCED DESIGN
- RECTANGULAR PASSAGE
- PARALLEL BASE
- JET AIR FLO
- VENTURI INTAKE PASSAGE
- COMPLETE WITH STUDS & GASKETS

ONLY $75.00

ALL - Cadillac	#3918
ALL - Oldsmobile	#3919
1953-56 Buick	#3922
1957-58 Buick	#3923

Water Outlet - Oldsmobile

| 1949-56 | #3920 | $12.00 |
| 1957-58 | #3921 | $12.00 |

"chev" pacesetters ☆ ☆ ☆ ☆ ☆ ☆ ☆ ☆

CHEVROLET V8
283 & 348   6 Carburetor

Two new Chevrolet manifolds - "With Heat", can be used from 2 to 6 carburetors - - "WORTH WAITING FOR" - - DIFFERENT

★ 283 ENGINE
AVAILABLE MAY 1st

#3924 $86.50
          Less Kit

★ 348 ENGINE
(IMPALA)
AVAILABLE JUNE 1st
$3925 $86.50
          Less Kit

2453

1097

AIR HORN ADAPTER, for installing air cleaners on 4-throat carburetors (all, except Holley or Lincoln) under low hoods . . . Catalog No. 2748

OFFIE-PLAC, durable thick metal, brilliant finish, trimmed in black, size 4-7/8" x 8-5/8" . . . . Catalog No. 2453

VALVE SPRINGS, heavy duty special intake valves made of top grade spring steel . . . . Catalog No. 1271

CARBURETOR LINKAGE - Positive action with no lost motion. Single and double linkages, rod lengths 3½ inch, 7 inch, 9 inch.

3½ inch Single . . . . .Catalog No. 1001
7 inch Single . . . . .Catalog No. 1002
9 inch Single . . . . .Catalog No. 1003
3½ inch Double . . . . .Catalog No. 1004
7 inch Double . . . . .Catalog No. 1005
9 inch Double . . . . Catalog No. 1006

Throttle Rod Sleeve – 3/16" Hole . . . . Catalog No. 1007
Throttle Rod Sleeve – 1/4" Hole . . . . Catalog No. 3498

FUEL BLOCKS, bright finish. Equalizes fuel pressure to all carburetors, dual, triple or four. Mounts on firewall of any make automobile; use flexline or neoprene hose to carburetor and fuel pump . .Cat. No. 1081

GENERATOR BRACKETS

Thru 1948 - Left hand straight . . . . Catalog No. 1093
            Right hand vertical . . . . Catalog No. 1094
1949 – 1953 - Right hand . . . . Catalog No. 1095
              Left hand . . . . Catalog No. 1096

FINNED COVER for Manifold Carburetor Pad . . . . . . Catalog No. 1097

ALL PURPOSE THROTTLE LINKAGE - Thru 1948

Rod Only . . . . Catalog No. 1098
Arm Complete . . . . Catalog No. 1099
Dual Linkage Complete . . . . Catalog No. 1100
Triple Linkage Complete . . . . Catalog No. 1173
4-Carburetor Linkage Complete . . . . Catalog No. 1174

ALL PURPOSE THROTTLE LINKAGE – 49-53. Specify Ford or Mercury and whether standard or automatic transmission.

Dual Linkage Complete . . . . Catalog No. 1175
Super Dual Linkage Complete . . . . Catalog No. 1176

BALL BEARING V8 FORD & MERCURY THROTTLE LINKAGE

For all Regular Duals, Adjustable . . . . Catalog No. 2864
For all Super Duals, Adjustable . . . . Catalog No. 2865
For all Triples, Adjustable . . . . Catalog No. 2866

EXHAUST PORT DIVIDERS - for Ford and Mercury 32-53 engines. Cast steel, comes complete with dowels for locking in place. Slight fitting necessary. Per set . . . . Catalog No. 1179

SPECIAL INTAKE MANIFOLD GASKETS – Dual purpose intake manifold gaskets made exactly the proper size for porting; may be used for template on block and manifold, then as gasket when finished. These are also available without intake port holes for blocks that are already ported. Will fit from 1932-53 Ford & Mercury . . . . . Catalog No. 1180

Without intake port holes . . . . Catalog No. 1181

MISCELLANEOUS

Fan Carrier Bracket . . . . Catalog No. 3499
Top Water Outlet . . . . Catalog No. 3496
Four Bolt Carb Cover Pad . . . . Catalog No. 3495
Fuel Log . . . . Catalog No. 3148
45° Elbow for Generator Clearance – 1949-53 . . . . . Catalog No. 3497
Oil Filter Bracket . . . . Catalog No. 3494
Special thick Intake Manifold Gaskets for
            Oldsmobile – 1949-53 . . . . . Catalog No. 3312
            Oldsmobile – 1954-56 . . . . . Catalog No. 3313

1380, 1181

1081   1179

1004, 1005, 1006

1001, 1002, 1003

1007

THRU 1948
A
Catalog No. 1093
B
Catalog No. 1094

1949 – 53

Right Hand
Cat. No. 1095

Left Hand
Cat. No. 1096

1098, 1099, 1100, 1173, 1174
2864, 2865, 2866

1271

3497

3496

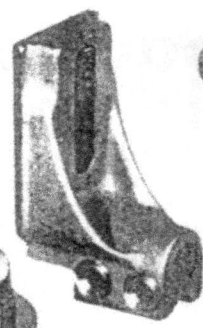

3499

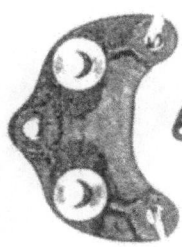

3494

3312
3313

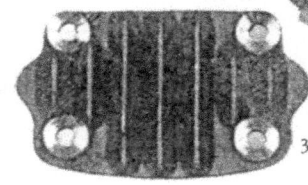

3495

3148

7

THE NEW OFFENHAUSER "2 BY 4" ADAPTER The ONLY adapter which allows you to mount two of the popular Stromberg or any two 3-bolt carburetors in place of EITHER an early OR a late quad carburetor. This universal "2 x 4" adapter fits any quad manifold. Can be used on Dual-Quad installations where the distance between centers of the Quad carburetors is at least 8-3/4". This is great for the enthusiast who wants to run on fuel using four easily-converted 3-bolt carburetors on a Dual-Quad manifold. Complete with studs and Allen Screws.

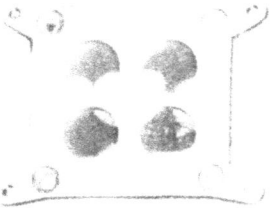

Catalog No. 3314

Now available also for wide base 1957-58 Ford, Lincoln and Mercury manifolds using 4 barrel carburetors. Cat. No. 3660

With this adapter you can install one of the improved-design widebase quad carburetors on your quad manifold. Get up-to-date performance, and additional hood clearance using the new, lower quad carburetors. (Price includes 4 Flat Head Machine Screws) Cat. No. 3311

Quad Leveling Block

This unit is a necessity when using a quad type carburetor on oval tracks. Made of aluminum, expertly machined, complete with studs, sets carburetor at approximately 8 degree angle.

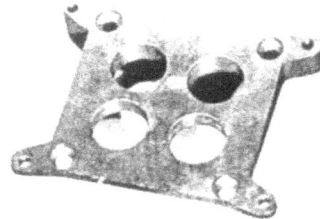

Cat. No. 3409

AIR HORN ADAPTER, for installing air cleaners on 4-throat carburetors (all, except Holley or Lincoln) under low hoods . . . Catalog No. 2748

FOR MARINE USE
CARBURETOR LEVELING BLOCK ASSEMBLY

Socket Head Allen screws, carburetor studs, nuts and washers furnished in kit.

Keeps carburetor at any desired angle from 0° to 12°.

Designed for engine in forward or reverse position.

Finest grade aluminum, expertly machined.

Fits Stromberg or Holley 3-bolt Ford type carburetor.

Be sure to specify whether front of engine is mounted towards front or rear of boat.

For Forward Position . . . Cat. No. 3124 - Price $8.25   Specify Degree
For Reverse Position . . . Cat. No. 3125 - Price $8.25   Desired.

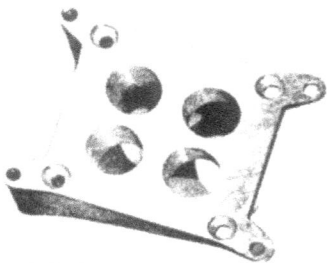

FOR MARINE USE

QUAD CARBURETOR LEVELING BLOCK ASSEMBLY

Type A - accomodates either wide or narrow base carburetor on manifold base of same type.
Cat. No. 3464

Type B - permits wide base carburetor to be installed on narrow base mounting.
Cat. No. 3465

MARINE USE

AVAILABLE MACHINED 12 OR 15 DEGREE ANGLE

Carb-Setter

An efficient, easy to handle tool that will hold most of the popular type carburetors in an upright position. You will be amazed how much faster you can repair and assemble a carburetor with this handy tool.

CARB SETTER

Part No. 3475

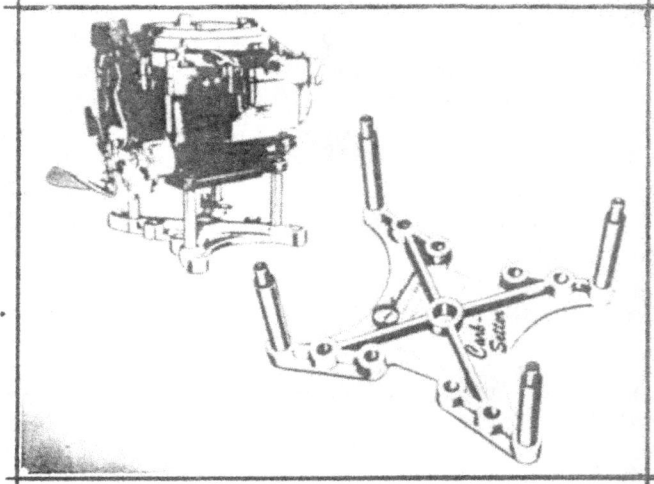

Main Cap Supports

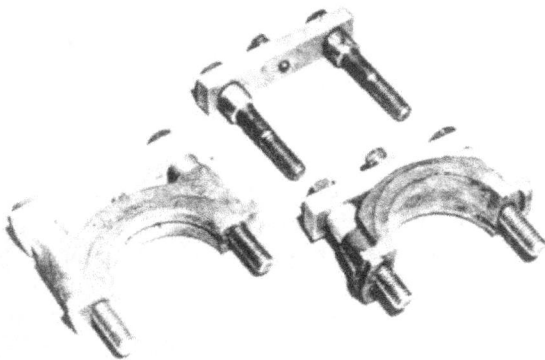

No longer necessary to install heavy Main Caps — these supports constructed of special steel complete with heat treated cap screws have been tested and proven on engines with outstanding HORSE POWER OUTPUT.

Ford & Mercury 59A Blocks
#3407 — $18.00
Ford & Mercury 8BA Blocks
#3408 — $18.00
Chevrolet OHV-V8—Complete Sets
#3650 — $21.50
Oldsmobile OHV-V8—Complete Sets
#3651 — $28.00
Pontiac OHV-V8—Complete Sets
#3652 — $28.00
Ford & Mercury OHV-V8—Sets
#3653 — $28.00

IGNITION LEAD PLATE

With adapter for 21A or 91A camshafts 1932 thru 1948 Ford and Mercury. For competition only — NOT RECOMMENDED FOR ROAD USE. When ordering, specify whether using with distributor or what type magneto also 91A or 21A camshaft.

Cat. No. 1091

Extra adapters specify 91A or 21A.

Cat. No. 1092

9

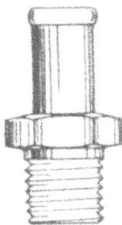 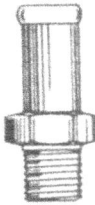 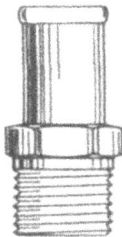 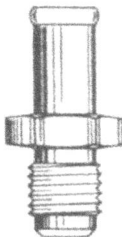 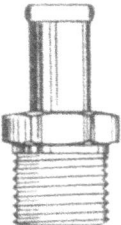

| Cat. No. | Cat. No. | Cat. No. | Cat. No. | Cat. No. |
| 2753 | 2752 | 2751 | 2689 | 2688 |

FUEL LINE FITTINGS

	Cat. No.
3/8" hose to Ford nose end	2689
3/8" hose to 1/8" pipe	2752
3/8" hose to 1/4" pipe	2753
3/8" hose to 3/8" pipe	2688
1/2" hose to 3/8" pipe	2751

Special Fitting for all Chrysler Products & Holly–Lincoln Carbs. to:

3/8" hose	3502
1/2" hose	3503

ADJUSTABLE CAM GEARS

FORD V–8 85 MERCURY AND FORD FERGUSON

For bolt on style cam. These gears enable you to juggle the cam timing 5 degrees either advance or retard.

No. 1273 21A .... $ 8.50
No. 1274 8BA .... $ 8.50

QUICK-CHANGE WELL PLUGS

FOR FORD STROMBERG CARBURETORS

Facilitates changing of main metering jets, with carbs on the manifold. Precision designed to permit maximum volume of fuel to flow in high speed metering systems.

No. 3850 ......$ .50 each
No. 3863 ......$ .50 each

HEX HEAD IDLE ADJUSTING NEEDLES

These little gems take the headache out of Multiple Carburetor Idle adjustments. For Stromberg Ford and Mercury Carbs.

NO. 3654—Stromberg Ford-Merc Carbs. $.35
NO. 3655—Chandler-Grove Holley $.35

ADJUSTABLE TAPPETS

JOHNSON

FORD, MERCURY AND FORD SIX
POSITIVE ACTION – ADJUSTABLE

NO. 1272   Set of 16 (std. length) $17.60

FUEL CONVERSION KITS

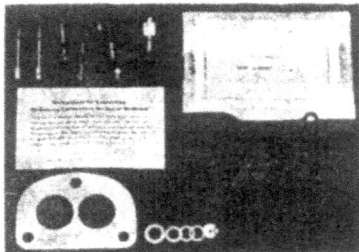

TO CONVERT STROMBERG CARBURETORS FOR USE WITH NITROMETHANE AND METHANOL FUELS

These kits are complete in every detail, with complete instructions covering mixing, handling, etc. Available for #97 and #48 models.

METHANOL ONLY
No. 1226   For #97 carbs. .... $5.50 EA.
No. 1227   For #48 carbs. .... $5.50 EA.

NITRO & METHANE
No. 3373   For #97 and #48 carbs. ... $5.50 EA.

NEEDLE SEAT FITTINGS

3852	1/8" NPT Female to 1/4" Hose for instruments	.45
3467	Ford Stromberg Carb. body to 3/8" hose	2.70
3468	Ford Stromberg Carb. body to 1/2" hose	2.70
3469	Ford Stromberg Carb. body* to 3/8" hose	3.00
3470	Ford Stromberg Carb. body* to 1/2" hose	3.00

*Carb. body must be drilled & tapped to 1/2"—20 SAE due to large oversize needle.

SPECIAL CARBURETOR SHAFTS

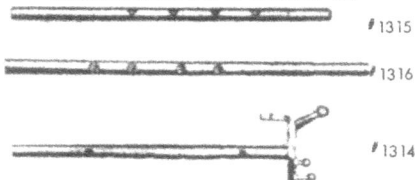

#1315
#1316
#1314

These are special long carburetor shafts made to accommodate special throttle linkage. Made for '56 Ford Holley, '55, '56, '57 Chevrolet Rochester, and the Holley 8BA-CG94.

No. 3599   1956 Ford Holley .... $ 1.60
No. 3601   1955-57 Chevy Rochester .... $ 1.90
No. 3600   8BA Stromberg .... $ 1.60

DISTRIBUTOR ADAPTER PLATES

FORD AND MERCURY (FLATHEAD)

Used when installing late model distributors on early model engines and early model front cover plate. Complete kit, including gaskets, screws.

No. 1332 ....$3.25

RAJAH IGNITION TERMINALS

Efficient spring type clip gives positive spark plug connection, plus easy removal for plug changing.

No. 1016   Straight type .... $ .60
No. 1015   Angle type .... $ .54

BALL BEARING THROTTLE LINKAGE

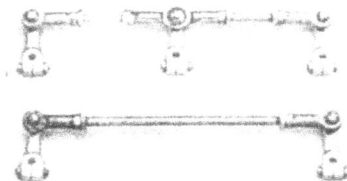

CAT. NO.

2866	Triple Linkage
2865	Super Dual Linkage
2864	Regular Dual Linkage
3851	Quad Linkage

Selective Throttle Linkage

For all triple and four carburetor manifolds overhead and flathead

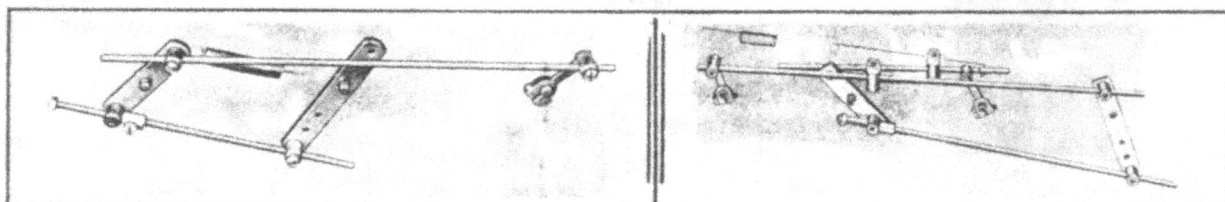

Stops all Automatic Transmission throttle problems – Once and for all! Operates as a single Carburetor to half throttle, then as triple. Can be installed as a Dual to half throttle, then as Triple. Can be changed in minutes to operate all 3 carburetors simultaneously. No Vacuum gimmicks – everything controlled by foot throttle. Saves fuel, smoother throttle, prevents jumpy starts, ends flat spots. Easy to install and adjust.

Standard Kit for Stromberg 48 or 97 Carbs.	$12.00	#3481
Chev. V8 w/Stromberg 48 or 97 w/2" extension	$15.00	#3482
Chev. V8 w/Rochester Carbs. w/long shafts	$16.50	#3483
Ford Holley or CG94 thru '55 w/long shafts	$16.00	#3865
Ford '56 or '57 Holley or Ford w/long shafts	$16.00	#3595
Throttle Shaft Extension – 2"	$ 1.00	#3484
Throttle Shaft Extension – 4"	$ 1.20	#3596

STANDARD SELECTIVE PROGRESSIVE 4-CARBURETOR LINKAGE

Eliminates overcarburetion at lower RPM's
Enables street use of competition engines
Provides startling gasoline economy . . . .
When used with 2 x 4 adapters and four 97's
it solves the dual quad problem.

Operates as dual Carbs. to half throttle –– then as four Carbs.
Can be set to operate four Carbs. at one time.
Unconditionally guaranteed when properly installed.

4 in line Carburetors, such as flat head, dual quads with 2 x 4 adapters Olds, Cad, Buick, Chrysler, DeSoto, Dodge	$18.00	#3597
4 staggered Carburetors – Chev.	$20.00	#3864
4 staggered Carburetors – Olds, Cad	$20.00	#3598

FUEL BLOCK
AND FUEL BLOCK KITS

Blocks   Hose
Kits   Clamps

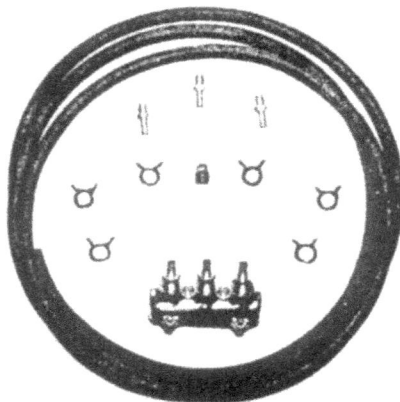

Due to the increased demand for Fuel Block Kits and their component parts, we are now offering a selection of hose types, clamps, etc.

FUEL BLOCK – Offenhauser polished fuel block. Adequate equalizer fuel well – supplies ample fuel equally to dual, triple or quad carburetors. Mounts on fire wall.   No. 1081

HOSE TYPE
- A – Red Neoprene – Highest quality Gates, double braid
- B – Black Neoprene – High quality, double braid
- C – Red Plastic – Medium quality
- D – Plastic clear – Medium grade
- E – TYGON – Extra High Grade, clear hose, 1/16" wall

Type	1/4" I.D. No.	1/4" I.D. Per ft.	3/8" I.D. No.	3/8" I.D. Per ft.	1/2" I.D. No.	1/2" I.D. Per ft.
A	3456	$ .40	3453	$ .42	3454	$ .60
B	3460	.42	2690	.42	3461	.60
C			2745	.30	2758	.42
D	3463	.38	3455	.42	3462	.60
E	3458	.40	3457	.70	3459	.80

CORBIN HOSE CLAMPS   Cat. No.

		Cat. No.
CS 6 – – – 3/8	Outside diameter hose	3438
CS 8 – – – 1/2	Outside diameter hose	3439
CS 9 – – – 9/16	Outside diameter hose	3868
CS10 – – – 5/8	Outside diameter hose	3440
CS11 – – – 11/16	Outside diameter hose	3869
CS12 – – – 3/4	Outside diameter hose	3501

Corbin clamps available up to 2-5/8" O.D.
Prices quoted on request.
Special pliers for Corbin clamps – – Net $2.50 each

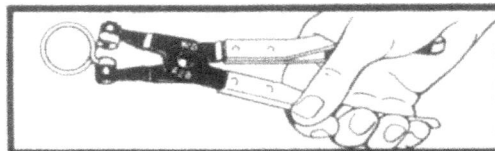

FUEL BLOCK KITS – –

Available now with a choice of hose listed complete with fuel block, fittings, carburetor fittings, etc., for dual, triple or quad carburetor.

Kit	TYPE HOSE A		B		C		D		E	
Dual	3441	9.50	3444	9.50	2733	9.00	3447	9.50	3450	10.50
Triple	3442	10.75	3445	10.75	2734	10.00	3448	10.75	3451	12.25
Quad	3443	12.25	3446	12.25	2735	11.50	3449	12.25	3452	14.25

HAND PRESSURE PUMPS

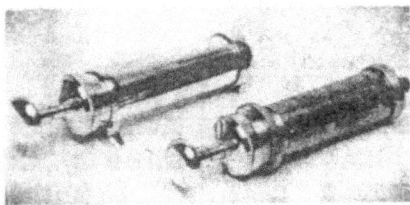

Large capacity hand pressure pumps. Made of 24ST Dural with heat treated ends. New type positive check valve. Available in either dash mount or side mount.

No. 1252 Side Mount $15.00

No 1253 Dash Mount $12.00

CARBURETOR ADAPTORS

NO. 3858 Ford 3-bolt to Chev. 2-bolt $ 2.95
NO. 3859 Ford 3-bolt to Dodge 2-bolt $ 2.95
NO. 3860 Ford 3-bolt to Merc. 4-bolt $ 3.25
NO. 3861 Ford 3-bolt to Quad (4BBL) $ 6.75
NO. 3862 Merc. 4-bolt to Quad (4BBL) $ 6.75

90 DEGREE FITTINGS

A Universal 90° fitting to fit 1/2"-20 needle seats. (Ford-Merc. Holley ~ some Rochester) Adaptable to other carburetors. Set hose end to desired position 90° from the carburetor, and tighten the bolt. Available for 3/8" or 1/2" hose.

No. 3505 3/8" ... $ 2.00
No. 3504 1/2" ... $ 2.00

BELL CRANK KITS

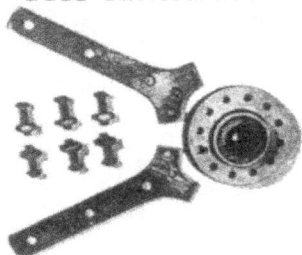

This precision made Bearing Assembly affords the user a choice of over 1,296 angles and leverage positions. Invaluable for custom engine installations in cars and boats. Uses the same 1-1/8" center line as selective linkage arms. Complete in kit form.

No. 3656 Bell Crank Kit . . . . $ 3.50

SPECIAL BELL CRANKS

By popular demand we offer the Extension Bell Crank so often needed for automatic transmission hook ups.

Number		List Price
3657	Chev. for use with selective linkage (Kit 333 CS) 1" extension	$3.00
3658	Chev. for use with Standard linkage 2" extension	3.00
3659	Olds - Cad - Hydra	3.00

HEX LOG FUEL BLOCKS

No. 3855 Triple (approx. 5" between carbs. Can be used for Dual) . . . . $ 6.00
No. 3856 Quad - Fits most 4 carb. manifolds $ 7.00
No. 3857 Sextet - Fits Log Type manifolds .. $ 7.00

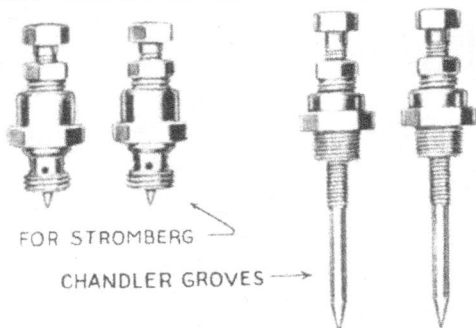

FOR STROMBERG

CHANDLER GROVES

ADJUSTABLE BRASS CARBURETOR JETS

Adjustable Jets for most all Ford and Merc carburetors. Effectively overcomes difficulties of varying atmospheric conditions, when fuel-air ratios change. Excellently manufactured of Brass. Gives mileage or power.

1225 Adjustable Jets, Ford Model 97 Stromberg . . . . . . per pair $ 3.50
1224 Adjustable Jets, Ford, Chandler Grove, Holley, per pair 3.50

6 AND 8 CARB LINKAGE AND HEX LOG FUEL BLOCKS

Patent Pending

Shown is a pair of No. 1460 Triple Hex Logs on a Log Manifold. Also #1458 Six Carb. Linkage.

No. 3853 Six Carb. Ball Bearing Linkage.... $44.00
No. 3854 Eight Carb. Ball Bearing Linkage . $50.00

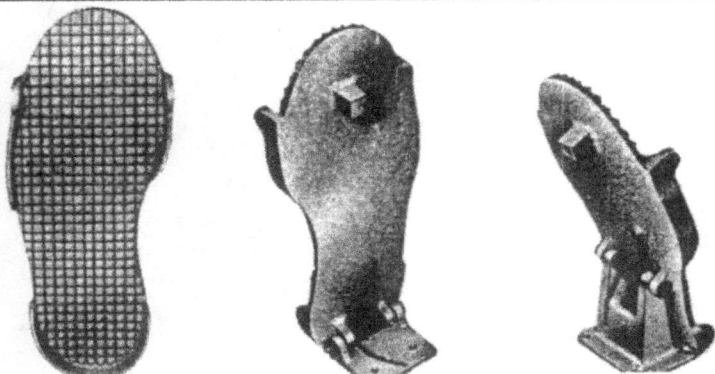

New improved Accelerator Pedal, spring loaded for automatic return, all aluminum, more bosses for connecting linkages. Available for either center pivot or heel pivot. Not only competition but for stock cars.

#3664 – Center pivot $6.75
#3665 – Heel pivot $6.75

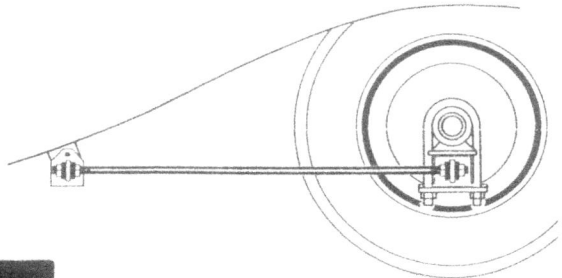
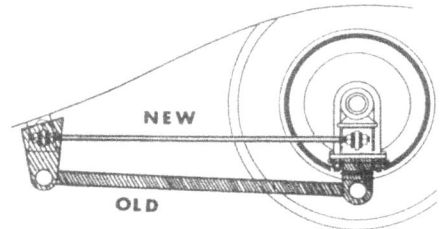

SENSATIONAL NEW ALUMINUM CLUTCH THATS ACTUALLY HARDER SURFACED THAN STEEL

Safety - serrated pins pressed then locked with cotter pins.

Special hardened steel bushings and pins in levers prolongs life, reduces wear.

Custom high tensile steel springs gives 2/3 more pressure to insure maximum performance.

Final assembly - completely adjusted with dial indicator to a tolerance of .003.

Light weight - 1/3 less than cast iron clutch.

Much better heat dissipation.

Safety - Tested at 9800 r.p.m.'s.

Surface harder than steel.

Pressure plate casting is case hardened by Sanford Process approximately .010 deep. As hard as sapphire.

COMPLETE CLUTCH COVER ASSEMBLY LESS DISC

CAR AND YEAR	TYPE	CLUTCH ONLY
Cadillac	11" C.F.	3514
Lincoln	11" C.F.	3515
Oldsmobile	11" C.F.	3516
Thunderbird	11" C.F.	3517
42—48 Ford and Mercury	10" C.F.	3518
49—50 Mercury	10" C.F.	3519
54—55 Ford	10" C.F.	3520
49—53 Ford	9½" C.F.	3521
35—42 Ford	9" C.F.	3522
Chrysler	10" B & B	3523
51—53 Mercury	10" B & B	3524
Chevrolet V8 with Corvette cover	10" B & B	3525-C
G.M.C.	10" B & B	3526
DeSoto	10" B & B	3527
Dodge	10" B & B	3528
Plymouth	10" B & B	3529
Early Plymouth	9" B & B	3530
Early Dodge	9" B & B	3531
Early Chrysler	9" B & B	3532
Early Studebaker	9" B & B	3533

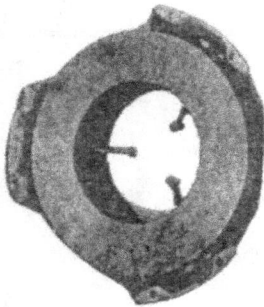

PRESSURE PLATES ONLY
EITHER C.F. OR BORG AND BECK

Many enthusiasts prefer building their own clutches. In this case we are now offering the plates only. This will enable your customers to have a clutch built to their exact specifications. Most of the clutch rebuilding firms can assemble these for you.

C.F. type

Borg & Beck type

CAR AND YEAR	TYPE	CAT. NO.	WEIGHT
Chev. V8 with Corvette cover	10" B.B.	3538-C	3 lb. 8 oz.
Cadillac, Lincoln, Oldsmobile, Thunderbird	11" C.F.	3534	4 lb. 9½ oz.
42—48 Ford, 42—50 Mercury, 54—55 Ford	10" C.F.	3535	3 lb. 11¼ oz.
49—53 Ford	9½" C.F.	3536	3 lb. 4 oz.
32—41 Ford	9" C.F.	3537	2 lb. 11¼ oz.
Chrysler Products, 51—52 Mercury	10" B.B.	3538	3 lb. 8 oz.
Early Chrysler and Studebaker	9" B.B.	3539	3 lb.

A completely NEW electronic tachometer

by **KELTRONIC INC.**

CATALOG NO. 3205

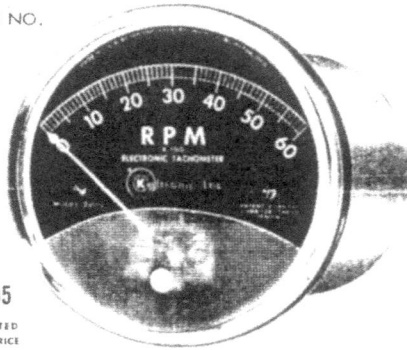

Drive by tachometer!
Be engine wise...economize!
Stop horsepower waste!
Save Gas!

R P M
ELECTRONIC TACHOMETER
Keltronic Inc.

$58 95
SUGGESTED
RETAIL PRICE

- 5 MINUTE INSTALLATION
- HIGHEST QUALITY EASILY READ DIAL
- FULL SCALE ACCURACY
- ALL ELECTRONIC COMPONENTS

NATIONAL DISTRIBUTORS:

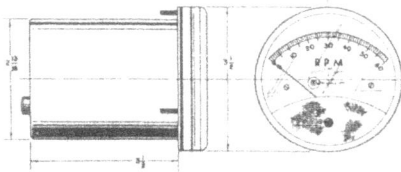

HOUSING: Spun aluminum container for automotive — Spun brass container for marine use.

METER: Clear lucite, black background, white numerals, easily read scale.

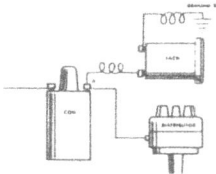

tachometer features

SIMPLE INSTALLATION
No special tools needed.

COMPLETELY SELF CONTAINED
No sending units, switches or cables required.

FULL SCALE ACCURACY
Accuracy not affected by length of leads, or direction of engine rotation.

BUILT-IN VOLTAGE REGULATOR
Accuracy not affected by voltage variation.

FOR AUTO-MARINE-INDUSTRIAL ENGINES
Battery ignition or magneto equipped (6 or 12 Volt Systems).

R.P.M. RANGES AVAILABLE
0-6000 R.P.M. Standard scale.
0-4000 R.P.M. and 0-8000 R.P.M. available.

INSTANTANEOUS READING
No needle oscillation.

MAGNETO IGNITION:

Hook up one wire from tachometer to primary stud of magneto; other wire from tachometer to a convenient ground.

IGNITION COIL AND DISTRIBUTOR:

Hook up wire from tachometer to point A as shown in diagram. Other wire from tachometer to a convenient ground; point B as shown in diagram.

ORDER CODE

1st number indicates number of cylinders.
2nd number indicates battery voltage.
3rd number indicates R.P.M. range.

MODELS AVAILABLE

4 x 6, 4 x 12, 6 x 6, 6 x 12, 8 x 6, 8 x 12
all models available in 0-4000, 0-6000, 0-8000 R.P.M.

	CYLINDERS		VOLTS		R.P.M.
EXAMPLE:	8	x	6	–	6

INFORMATION REQUIRED WHEN ORDERING KELTRONIC TACHOMETERS:

1. Make of engine
2. Number of cylinders
3. Battery Voltage
4. Type of ignition —
   (if other than standard).

YOUR REGULAR
DISCOUNT APPLIES

Special light for tachometer – – #3500 $ 2.50

Marine type tachometer with brass case – –
                                              #3540 $61.95

ENGINE ADAPTERS

FOR

O.H.V. V-8 ENGINES

install a big powerful overhead valve V8 engine the easy way - with an engine adapter

ALL THE FAMOUS MAKES

Cragar, Hildebrandt, Wilcap

Put extra punch in any Ford or Mercury

- HEAT TREATED ALUMINUM
- IMPROVED DESIGN
- PRECISION MACHINING
- GUARANTEED FIT

LATEST TYPE, LOCATING PILOT BEARING ADAPTERS

The following engines can be fitted to '32 to '48 Ford and '39 to '50 Mercury transmissions. Price includes pilot bearing adapter and instructions.

NO. 309	Buick, '53-'56	$55.75
NO. 3702	Buick, '57-'58	$55.75
NO. 303	Cadillac, '55-'58	$64.50
NO. 301	Cadillac, '49-'54, Oldsmobile '49-'56	$31.50
NO. 307	Chevrolet OHV V8, '55-'58	$59.50
NO. 401	Chrysler '54-'58, DeSoto '52-'58, Dodge '53-'57, Plymouth '55-'58	$74.50
NO. 403	Chrysler, '51-'53	$53.50
NO. 600	Ford '54-'57, Mercury '54-'57, Also 292 cu.in. Ford '58 and 223 (6 cyl.) Ford '54-'58	$59.50
NO. 3635	Packard, Nash, Hudson, '55-'57	$35.00
NO. 3606	Pontiac 57-58 OHV GV8	$31.50

The following engines can be fitted to '49 to '57 Ford and '51 to '57 Mercury transmissions. Price includes pilot bearing adapter and instructions.

*NO. 310	Buick, '53-'56	$67.50
NO. 302	Cadillac, '55-'58	$77.75
NO. 304	Cadillac '49-'54, Oldsmobile '49-'58	$53.50
NO. 3634	Chevrolet OHV V8, '55-'58	$39.50
*NO. 402	Chrysler '54-'58, DeSoto '52-'58, Dodge '53-'58, Plymouth '55-'58	$81.75
NO. 3704	Lincoln, '52-'57	$69.00

Pilot bearing adapters may be ordered separately.

300-1—Cadillac and Oldsmobile...$4.25

400-1—All Chrysler Products...$4.25

NO. 600
FORD OHV V8, 54–58
MERCURY OHV V8, 54–58
TO EARLY F. AND M.
TRANSMISSIONS

NO. 302
CADILLAC 55–58
TO LATE F. AND M.
TRANSMISSIONS

NO. 301
CADILLAC 49–54
OLDSMOBILE 49–58
TO EARLY F. AND M.
TRANSMISSIONS

NO. 303
CADILLAC 55–58
TO EARLY F. AND M.
TRANSMISSIONS

NO. 304
CADILLAC 49–54
OLDSMOBILE 49–58
TO LATE F. AND M.
TRANSMISSIONS

NO. 3635
1955–57 PACKARD,
NASH, HUDSON
OHV V8
TO EARLY F. AND M.
TRANSMISSIONS

NO. 307
55–58 CHEVROLET
OHV V8
TO EARLY F. AND M.
TRANSMISSIONS

NO. 3634
55–58 CHEVROLET
OHV V8 TO LATE
49–57 FORD AND 51–57
MERC. STEEL PLATE
AND PILOT ADAPTER

NO. 309
BUICK 53–56
TO EARLY F. AND M.
TRANSMISSIONS

NO. 3606
57–58 PONTIAC OHV V8
TO EARLY F. AND M.
TRANSMISSIONS

NO. 3704
LINCOLN 1952–57
TO 1949–57 FORD
AND 1951–57 MERCURY
TRANSMISSIONS

NO. 403
CHRYSLER 51–53
TO EARLY F. AND M.
TRANSMISSIONS

NO. 401
CHRYSLER 54–58
DeSOTO 52–58
DODGE 53–58
PLYMOUTH 55–58
TO EARLY F. AND M.
TRANSMISSIONS

OLDSMOBILE STARTER ADAPTER

Solving the problem of "where shall I put the starter, the steering is in the way?" A common problem with Olds engines in early Ford and other chassis. The new Hildebrandt adaptor allows switching of the starter motor to the opposite side of the engine. Utilizes all original bolts and holes. Retains original strength, at the same time reducing weight 150%. Cast and machined of aluminum, adaptor fits all Olds OHV V8 engines. From there the above adaptors may be used for Ford transmissions, etc.

NO. 3605 Oldsmobile Starter Adapter $37.00

TRANSMISSION ADAPTERS

LA SALLE-CAD. TRANSMISSION ADAPTER

New Hildebrandt engine adaptor. Cast aluminum, sand blast finish. Has Early Ford mounting flange or bolt pattern. Adapts to any other adaptor that has Early Ford Trans. Flange. From there to Cad. or LaSalle trans. Uses stock Early Ford Clutch Release, shaft, fork, throw out bearings, etc. This allows the use of the mentioned transmissions in place of the Early Ford type. Used after the installation has been made with an Early Ford trans., and same does not stand up. When using this adaptor, no clutch linkage changes are necessary, such as pendalum pedals, slave cylinder, etc.

NO. 3640 Hildebrandt engine adaptor $59.50

PACKARD TRANSMISSION ADAPTERS

Same as 3640 except it fits Packard transmission
NO. 3641 . . . . . . $59.50

BUICK TRANSMISSION ADAPTERS

(not shown)

Same as 3640 except it fits Buick Roadmaster transmission
NO. 3648 . . . . $59.50

CAD-LASALLE TO 55-58 CHEVY ADAPTER PLATES

To adapt Cad.-LaSalle transmission to 55-58 Chev.
NO. 3705 . . . . . . $40.00

Offenhauser

O.H.V. V-8s BALANCED AND STROKED CRANK ASSEMBLIES

Balanced stroker assemblies for OHV V8s include the following new parts:

Crank with throws indexed and polished (new or used)
Jahns Pistons pin fitted
Grant rings
New rods balanced end for end and
　magnafluxe inspected - Journal size std.
　bearing inserts are furnished -
(Balancing uses weight of stock inserts)
(Mains not included - .010 mains)

SPECIFY NEW OR USED CRANK ASSEMBLY

Engine	Stroke with any bore	Price with used crank	Price with new crank
Buick	3.450	$440.00	$494.00
Cadillac 49-54	3-7/8	$405.00	$459.00
Cadillac 49-54	4	$441.45	$506.25
Cadillac 55-57	3-7/8	$421.20	$476.55
Chevrolet 55-57	3-1/4	$311.85	$379.35
Chevrolet 55-57	3-3/8	$355.05	$422.55
Chevrolet 55-57	3-1/2	$386.10	$452.25
Chevrolet 58	3-1/2	xxx	$423.90
Chevrolet 58	3-5/8	xxx	$467.10
Chevrolet 58	3-3/4	xxx	$496.80
Chrysler 51-57	3-7/8	$396.90	$477.90
Chrysler 51-57	4	$427.95	$508.95
Chrysler 51-57	4-1/8	$457.65	$540.00
Ford 55-57	3.500	$324.00	$400.95
Ford 55-57	3.640	$368.55	$427.95
Ford 58	3-3/4	xxx	$445.50
Ford 58	3-7/8	xxx	$476.55
Ford 58	4	xxx	$507.60
Mercury 55	3.500	$324.00	$400.95
Mercury 55	3.640	$368.55	$427.95
Mercury 56-57	3.640	$341.55	$400.95
Oldsmobile 49-56	3-11/16	$359.10	$405.00
Oldsmobile 49-56	3-13/16	$379.35	$436.05
Oldsmobile 49-56	3-15/16	$419.85	$467.10
Oldsmobile 57	3-15/16	$400.95	$436.05
Oldsmobile 57	4-1/16	$430.65	$465.75
Oldsmobile 57	4-3/16	$461.70	$496.80
Pontiac 55-57	3-1/2	$415.00	$450.00

*These were estimated strokes and not confirmed.
Specify exact make and model of cylinder blocks on all orders.

Note: Olds 1957 Assembly must be balanced with flywheel and damper. You must furnish flywheel. Prices above include damper but not flywheel.

SPECIAL ROD BEARINGS
ADD PRICE TO ABOVE ASSEMBLY PRICES

NO. 2436　FORD, MERCURY, THUNDERBIRD --
　　Rods reworked for 99A bearings, including bearings $40.50
NO. 2437　BUICK, CADILLAC, OLDSMOBILE & 51-56 CHRYSLER--
　　Rods reworked for full floating, including bearings $76.50

NOTE:　Large diameter cams cannot be used on stroked crank O.H. V. V8 engines. Even with average size cams some con. rod bolt heads hit the cam when increasing the stroke more than 1/4"

Another problem with extra long strokes, especially when pistons are heavy, is that plates have to be welded inside the counter-balances. If you burn a bearing bad, crank throws cannot be reground without removing these plates.

DELIVERY - ALLOW AT LEAST 15 DAYS FOR DELIVERY AS EACH ASSEMBLY IS CUSTOM MADE TO YOUR SPECS.

FULLY COUNTER-BALANCED CHEVY V8 CRANKSHAFTS - 1955-57

STOPS FLEXING AT HIGH R.P.M.

Special for Chevy V8's 1955-57. This crank will allow much more RPM's to be turned besides making the engine much smoother. Add $64.80 to the Balanced Assembly prices shown at top of this page, and specify.

FULL FLOW OIL FILTERS

Cast of aluminum, this full flow filter is recommended to be used on all OHV V8 engines. Uses late Ford or Mercury oil filter element. Designed to be mounted in any convenient location. Used in conjunction with all oil filter by-pass plates.
NO. 3637　Full Flow Oil Filter $24.50

OIL BY-PASS UNITS
OHV V8

CHRY. PRODUCTS　　FORD-MERC-LINCOLN　　OLDSMOBILE V8　　CHEVY V8

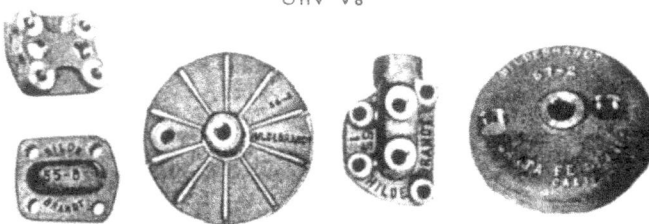

When switching starter to the right side of the engine, the oil filter has to be removed. If no oil filter is to be used, the Ford oil pressure sending unit may be installed in either the "In" or "Out" openings of the By-Pass Plate. Olds uses stock spring and cap inside by-pass.

NO. 3604　Oldsmobile Unit .... $ 9.50
NO. 3644　Chrysler, DeSoto, Dodge, Plymouth (block-off) .... $ 5.00
NO. 3645　Chrysler, DeSoto, Dodge, Plymouth (by-pass) .... $ 5.50
NO. 3639　Ford, Mercury and Lincoln .... $ 7.50
NO. 3646　Chevrolet V8 1955-57 .... $ 9.50
NO. 3647　Buick V8 (not shown) .... $ 9.50

WIL-CAP
ALUMINUM FLYWHEELS
FOR ALL OHV V-8 ENGINES

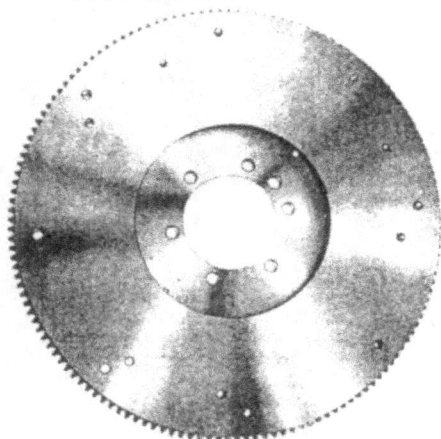

Cast only of the finest aluminum and heat treated ●● Stock flywheel dimensions ●● Only one-third the weight of stock flywheel ●● Machined to the closest tolerances ●● Drilled for stock clutch, 10" Ford Long and also 11" Ford clutch ●● On special order can be drilled for any clutch. Most stock clutches are 10" Borg & Beck.

BE SURE TO SPECIFY YOUR USE

WIL-CAP FLYWHEELS ARE COATED BY THE REVOLUTIONARY SANFORD PROCESS
• Coatings are electrochemically produced resulting in a composition of amorphous alumina, an extremely hard substance • Stable at temperatures in excess of melting point of aluminum • Not only adheres to, but is a part of the flywheel, will not chip, peel or flake • Hardness as shown by scratch test is between 7 and 9 on MOHS scale • Sanford hardened surfaces do not have a calculable effect on fatigue life • Results . . A very hard, abrasive resistant surface.

Tested and approved for the following engines.

Buick OHV V8 52-56
Buick 57—58
(use with Ford 11" clutch only)
Cadillac OHV V8 49-53
Cadillac 54—58
Oldsmobile OHV V8 49-53
Oldsmobile 54—58
GMC (big) OHV V8 49-53

Chevrolet OHV V8 55-57
drilled for 10" stock,
11" stock and 11" Ford truck
Chevrolet 55—58
Including 348 cu.in. engine

Chrysler OHV V8 33-58
DeSoto OHV V8 33-58

Dodge OHV V8 33-58
Plymouth OHV V8 33-58

Ford 54-57
Mercury 54-57
Thunderbird 54-57
Turnpike Cruiser 58
Lincoln OHV V8 52-57
Pontiac 55-57
GMC (small) OHV V8
Studebaker OHV V8 All
(not Packard)

$59.50

STATE MAKE AND MODEL
WHEN ORDERING Part No. 3867

FAMOUS
AUBURN
CLUTCHES

Specially designed, brand new, matched and balanced sets. Approximately two pounds lighter than a stock assembly. Allows easy shifting at high RPM. For the Ford, Mercury, Chevy, and G.M.C. Outright sale only.

Cat. No.	Application	Size of Spline	Price
NO. 3842	Chevrolet 9" (woven lining) 6 cyl.	1-1/8"x10	$29.25
NO. 3843	Chevrolet 9" (molded lining) 6 cyl.	1-1/8"x10	$29.25
NO. 3844	G.M.C. 11" 6 cyl.	1-1/8"x10	$45.25
NO. 3845	Ford 9" Long (woven lining)	1-3/8"x10	$31.90
NO. 3846	Ford 9" Long (molded lining)	1-3/8"x10	$31.75
NO. 3847	Ford 11" Long (woven extra heavy duty)	1-3/8"x10	$33.10
NO. 3848	Mercury 10" Long (woven lining) (also Ford)	1-3/8"x10	$30.00
NO. 3849	Mercury 10" Long (molded lining)(also Ford)	1-3/8"x10	$25.40

SCHIEFER ALL NEW ALUMINUM FLYWHEELS
BONDED BRONZE AND STEEL FACE

(ELECTRONICALLY BALANCED)

The Schiefer all new aluminum flywheels weigh 10-1/2 to 11-1/2 lbs. Casting: 356T6 Permanent Mold. Strength: 48,000 P.S.I.

Schiefer Flywheels are cast from aircraft-type aluminum alloy and are heavily ribbed for maximum strength. Will not blow up under extremes of temperature and stress. The exclusive Schiefer "bonded face" assures full-contact clutching action and longer clutch life than any other lightweight flywheel.

Threaded steel inserts for pressure plate screws provide positive alignment, eliminate thread stripping. Schiefer light weight flywheels are a favorite with the record holders of drag and track and have stood up under the most severe tests!

MODELS ARE AVAILABLE FOR MOST MAKES OF ENGINES
BUTTON STYLE (NO GEAR) ON SPECIAL ORDER

Cat. No.	Model and Year	Clutch Cover	Price
No. 3822	52-56 Buick V8	11" Ford long	$ 66.00
No. 3823	57-58 Buick V8 (incl. Starter Spacer Adapter)	11" Ford long	70.00
No. 3436	49-53 Cadillac	11" Cad.-Olds.-Ford long	60.00
No. 3437	54 thru early 56 Cadillac	11" Cad.-Olds.-Ford long	63.50
No. 3824	Late 56-58 Cadillac	11" Cad.-Olds.-Ford long	63.50
No. 3422	42-53 Chev.	Stock 9-1/8 Chev.	58.50
No. 3425	54-58 Chev. 6	Stock 10" Chev.	58.50
No.3428	55-58 Chev. with auto-trans.	Stock 10" B&B & 10-1/2"B&B	58.80
No.3829	V8 Corv. & with overdrive	Stock 11" B&B	61.80
No.3830	Power Pack	10" Ford long	58.80
No. 3831	55-58 Chev. V8 eng. conv. & Ford trans. etc.	11" Ford long	60.00
No. 3832	57-58 Chrysler	11" 8 bolt B&B	60.00
No. 3429	33-58 Chry.-DeSoto-Dodge-Ply.	9-1/4" & 10" B&B	58.00
No. 3431	33-58 Chry.-DeSoto-Dodge-Ply.	10" Ford long	57.00
No. 3430	33-58 Chry.-DeSoto-Dodge-Ply.	11" Ford long	59.00
No. 3833	57-58 Plymouth Fury	10-1/2" B&B	59.50
No. 3834	58 Edsel	11" Ford long	66.00
No. 1196	32-48 Ford V8 - Mercury	10" Ford long	52.30
No. 1197	49-53 Ford V8 - Mercury	9-1/2" long & 10" B&B	53.00
No. 3433	49-53 Ford-Mercury	10" Ford long	52.30
No. 3434	54-57 Ford-Mercury *	10" long & 10" B&B	58.50
No. 3435	54-57 Ford-Mercury *	11" Ford long	59÷50
	*Thunderbird & OHV Ford 6 58 Ford (205 & 292 eng. only) not 58 T'bird		
No. 3835	58 Ford (332 & 352 eng. & T'bird)	11" Ford long	66.00
No. 3836	52-57 Lincoln-F8-Turnpike Cruiser	11" Ford long	66.00
No. 3837	58 Lincoln	11" Ford long	66.00
No. 3838	57 Mercury Turnpike Cruiser	11" Ford long	66.00
No. 3839	58 Mercury	11" Ford long	66.00
No. 3436	49-53 Oldsmobile	11" Cad.-Olds.-Ford long	60.00
No. 3437	54-58 Oldsmobile	11" Cad.-Olds.-Ford long	63.50
No. 3825	55-56 Packard-Hudson-Nash (not Rambler)	11" Ford long	70.00
No. 3826	55-57 Pontiac-Small GMC V8	10" B&B	60.00
No. 3827	55-58 Pontiac-Small GMC V8	10" Ford long	60.00
No. 3828	55-58 Pontiac-Small GMC V8 Eng.conv.only	11" Ford long	63.50
No. 3840	58 Pontiac	11" B&B	63.50

PONTIAC V8 - CHEVROLET V8
THROW-OUT BEARING
ADAPTOR KIT

This adaptor enables a 10" Ford long type clutch to be installed with standard shift transmissions.

Schiefer Conversion Kits are available for installing long, Schiefer Ford-type 10" clutch on redrilled stock flywheels or aluminum flywheels drilled for 10" long type clutch. Standard clutch unit must be of the finger type and not the diaphram style -- if it is not, then the clutch housing must be changed to style used with finger type.

No. 3841   10" Ford long $20.00

HEDMAN HEDDERS

180 DEGREE FIRING ORDER
HEDMAN HEDDERS
TUNED DESIGN

FORD THUNDERBIRD "HEDDER"

The new design differs in many ways from the previous design exhaust "Hedders". In the former design, the collector or log type was used. However, it is a known fact that best results are obtained when all exhaust ports are of an equal length so one port does not fire past another port at its firing time which causes restrictions that are characteristic of the stock cast iron manifolds. In "Hedders", the principle of vacuum or scavenging is used. By this we mean the exhaust gases, or pressure emitted by one cylinder after firing, coming through its "Hedder" port must pass another cylinder "Hedder" port thus creating a vacuum action. This forms a low pressure area in the scavenged port which permits a full intake charge to be drawn in during the overlap period of valve timing.

This "Hedder" is designed for the Thunderbird chassis only. It will not fit other cars regardless of chassis. See listing for passenger cars.

EXHAUST HEDDERS
(Includes all necessary bolts, gaskets, etc.)

FORD V8	Stock No.	Price
1955-57 All, except Thunderbird	H-FM1	$60.50
1955-57 Thunderbird	H-FT1	$60.50
1954-58 F-1 & F-100 Pickup Truck	H-FO1	$60.50
1958 All, except 332 - 352 cu. in. engine	H-FM1	$60.50
MERCURY		
1955-57 All Models	H-FM1	$60.50

EXTENSIONS TO MUFFLERS
(Includes all necessary clamps, gaskets, etc.)

FORD V8		
1955-56 All, except Thunderbird & without factory duals	X-FO2	$ 5.25
1955-56 All, except Thunderbird & with factory duals**	xxxxx	xxxxx
1955-57 Thunderbird	X-FT1	$ 6.00
1954-58 F-1 & F-100 Pickup Trucks	X-FO1A	$ 6.00
1957 All, with factory duals, except Thunderbird**	xxxxx	xxxxx
1957 All, without factory duals except T'bird & conv't	X-FO3	$ 4.35
1957 Convertible, without factory duals	X-FO4	$ 4.35
1958 All, except 332 - 352 cu. in. engine	X-FO3	$ 4.35
MERCURY		
1955-57 All, without factory duals	X-ME1	$ 5.25
1955-57 All, with factory duals**	xxxxx	xxxxx

**None required, Hedders connect directly to standard exhaust pipes.

CUSTOM "HEDDERS"

	Stock No.	Price
1955-57 Chevrolet all models	H-CHU1	$62.50
1958 Corvette only	H-CHU2	$62.50
1955-57 Ford, Mercury, Thunderbird	H-FOU1	$62.50

extensions not available, must be custom made.

PLYMOUTH V8 "HEDDER"

EXHAUST HEDDERS

	Stock No.	Price
1957-58 All except Convertible & Golden Commando eng.	H-PL1	$62.00

EXHAUST EXTENSIONS

	Stock No.	Price
1957-58 All, except Convertible & Golden Commando eng. and without factory duals	X-PL1	$ 9.30
1957-58 All, except Convertible & Golden Commando eng. and with factory duals	X-PL2	$ 9.30

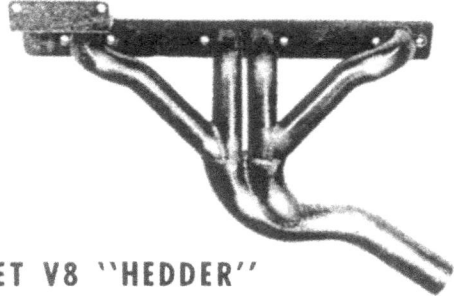

CHEVROLET V8 "HEDDER"

WILL FIT ALL MODELS
INCLUDING TRUCKS

EXHAUST HEDDERS
(Includes all necessary bolts, gaskets, etc.)

CHEVROLET V8	Stock No.	Price
1955-57 All, except Corvette	H-CH1	$62.50
1955-57 Corvette	H-CH2A	$62.50
1955-58 3100 Pickup Trucks 1/2 ton	H-CH1	$62.50
1958 All, except Corvette & without 348 engine	H-CH3	$69.00
1958 Corvette	H-CH4	$63.50
1958 All, except Corvette & with 348 engine	H-CH5	$70.50

EXTENSIONS TO MUFFLERS
(Includes all necessary clamps, gaskets, etc.)

CHEVROLET V8		
1955-57 All, except Conv't. & Corvette	X-CH1	$ 5.25
1955-57 Convertible	X-CH2	$ 6.25
1955-57 Corvette	X-CH4	$ 5.25
1955-58 3100 Pickup Trucks 1/2 ton	X-CH3	$ 6.50
1958 All, except Corvette & without 348 engine	X-CH5	$ 5.25
1958 Corvette	X-CH4	$ 5.25
1958 All, except Corvette & with 348 engine	X-CH5	$ 5.25

MOTOR MOUNTS

FRONT MOTOR MOUNTS

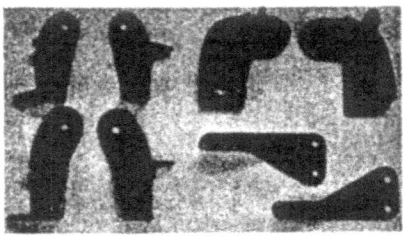

INSTALLS LATE MODEL V8'S INTO '37—'53 FORDS AND MERCS

NO OTHER PARTS ARE REQUIRED BOLTS TO ORIGINAL STOCK MOUNTS ALSO ADAPTS TO '32—'36 FORDS

Complete with bolts as required for special installations.

Mounts are made from heavy steel electrically welded. Olds mounts include special bolts needed for installation. All mounts are finished grey. Instructions are included in mounts where installation is not obvious. The mounts are boxed in pairs and weight is approx. 6 lbs.

CHASSIS	ENGINES						
	BUICK	CAD	CHRY.	CHEVY	DESOTO	FORD-M	OLDS
1937-48 Ford-Merc.	x x	No 3678 $17.50	No 3682 $17.50	No 3686 $15.50	No 3690 $17.50	No 3694 $28.75	No 3698 $17.50
1949-53 FORD	No 3676 $17.50	No 3679 $17.50	No 3683 $17.50	No 3687 $12.50	No 3691 $17.50	No 3695 $19.50	No 3699 $17.50
1949-50 MERCURY	x x	No 3680 $19.50	No 3684 $19.50	No 3688 $19.50	No 3692 $17.50	No 3696 $28.75	No 3700 $19.50
1951-53 MERCURY	No 3677 $17.50	No 3681 $19.50	No 3685 $19.50	No 3689 $19.50	No 3693 $17.50	No 3697 $19.50	No 3701 $19.50

FINISHED   PINFIT   WITH PINS

NOTE: The pistons listed below are primarily for drag or racing engines. Be careful that you don't pick too high a compression piston if you intend using gasoline for fuel. The same care should apply when high percentage of nitro is contemplated. The more nitro used the lower the compression ratio should be. Methanol alone can use as high a compression ratio as your engine combustion chamber design will allow.

FOR PISTONS FURNISHED IN OTHER STROKES THAN SHOWN
IN CATALOG ADD — LIST PRICE EACH — $1.50

FOR ALL OTHER PISTONS FINISHED TO A SIZE NOT SHOWN
IN CATALOG ADD — LIST PRICE EACH — $1.20

BUICK	$104.80
1957-58	4-1/8
50-60-70	4-3/16
10 & 11.5 comp. ratio	4-1/4
specify	

BUICK	$95.20
1956	4-1/8
50-60-70 (std. bore is 4")	4-3/16

BUICK	$98.40
1954-55	3-5/8
10.0:1 Comp. ratio	3-3/4
11.0:1 Comp. ratio	3-13/16
40 series Specify ratio	

BUICK V8	$95.20
1954-55 (std. bore is 4")	4-1/8
50-60-70	4-3/16

CADILLAC	$82.40
1953-58	3-13/16
Flat Top	3-7/8
Comp. ratio depends	3-15/16
on heads used plus	4
increase due to bore	4-1/16
	4-1/8

CADILLAC	$88.00
1953-58	3-13/16
Domed High Comp.	3-7/8
Increases comp. by	3-15/16
1 ratio over ratio	4
of heads used plus	4-1/16
increase due to bore	4-1/8

CADILLAC	$82.40
1949-52	3-13/16
Flat Top	3-7/8
Ratio depends on heads	3-15/16
	4

CHEVROLET	$92.00
1958 348 Eng. Domed	4-1/8
10.1 comp. ratio	4-1/4
12.1 comp. ratio specify	

CHEVROLET V8	$82.40
1955-58 (except 348 eng.)	3-3/4
Flat Top Deep valve relief	3-7/8
relief. Can be used	3-15/16
with any heads.	4

CHEVROLET V8	$86.40
1955-58 (except 348 eng.)	3-3/4
Flat Top Deep valve	3-7/8
relief. 1/8 l.c. for 1/4	3-15/16
stroker. Can be used	4
with any heads.	

CHEVROLET V8	$86.40
1956-58 (except 348 eng.)	3-3/4
Domed Hi comp. For	3-7/8
power pack heads only.	3-15/16
Deep valve relief.	4
10.1 comp. ratio	

CHEVROLET V8	$86.40
1955-58 (except 348 eng.)	3-3/4
Domed Hi comp. Can be	3-7/8
used with all heads ex-	3-15/16
cept power packs.	4
Deep valve relief.	
10.1 comp ratio	

CHRYSLER V8	$96.80
1957-58	4
Domed High comp.	4-1/16
10.5 to 1 comp. ratio	4-1/8
12 to 1 comp. ratio specify.	

CHRYSLER V8	$96.80
1955-56 Domed	3-13/16
9.0:1 comp. ratio	3-7/8
10.0:1 comp. ratio	3-15/16
specify ratio	4

CHRYSLER V8	$82.40
1951-56	3-13/16
1/8 l.c. for 3-7/8 stroke	3-7/8
1/4 l.c. for 4-1/8 stroke	3-15/16
specify stroke	4

CHRYSLER V8	$96.80
1951-54 Domed	3-13/16
9.0:1 comp. ratio	3-7/8
10.0:1 comp. ratio	3-15/16
specify ratio	4

DESOTO V8	$96.80
1956	3-23/32
	3-3/4

DESOTO V8	$96.80
1956 Domed	3-23/32
11.75:1 comp. ratio	3-3/4
10.0:1 comp.ratio specify	

DESOTO V8	$96.80
1952-55	3-5/8
	3-23/32
	3-3/4

DESOTO V8	$96.80
1952-55 Domed	3-5/8
10.5:1 comp. ratio	3-23/32
9.0:1 comp. ratio specify	3-3/4

DODGE	$95.20
1956	3-5/8
D-500	3-3/4

DODGE V8	$96.80
1953-56	3-7/16
	3-9/16
	3-5/8
	3-11/16
	3-3/4

DODGE V8	$96.80
1953-56	3-7/16
Domed	3-9/16
High comp.	3-5/8
9.0:1 comp. ratio	3-11/16
	3-3/4

EDSEL	$100.00
1958 Corsair & Citation	4.200

FORD V8	$77.60
1955-56-57	3-5/8
1958 292 engine	3-3/4
	3-7/8

FORD V8	$82.40
1955-56-57	3-5/8
1958 292 engine	3-3/4
9.0:1 comp. ratio	3-7/8
10.5:1 comp. ratio specify	

FORD V8	$84.00
1955-56-57 3.5 stroke	3-5/8
1958 292 engine	3-3/4
.200 stroker - 100 LC	3-7/8

FORD V8	$77.60
1954	3-5/8

LINCOLN	$86.40
1956-58	4
& 368 engine	4-1/16
	4-1/8

LINCOLN V8	$82.40
1952-55	3.800
	3-15/16
	4

MERCURY	$80.00
NOTE 3.925 BORE SIZE IS	$89.60
1956-57	3.800
Thunderbird V-8	3-7/8
312 engine	3.925

MERCURY	$84.80
NOTE 3.925 BORE SIZE IS	$94.40
1956-57 T'bird V8 312	3.800
9.01 comp. ratio	3-7/8
10.5:1 comp. ratio specify	3.925

MERCURY	$89.50
NOTE 3.925 BORE SIZE IS	$101.50
1956-57 T'bird V8 312	3.800
3.640 stroke	3-7/8
.200 stroker = 100 LC	3.925

MERCURY V8	$77.60
1955 Thunderbird V8	3-3/4
(std. bore 3-3/4)	3-7/8

MERCURY V8	$82.40
1955 Thunderbird V8	3-3/4
9.0:1 comp. ratio	3-7/8
10.5:1 comp. ratio specify	

MERCURY V8	$82.40
1955 Thunderbird V8	3-3/4
(std. bore 3-3/4)	3-7/8
3.5 stroke	
.200 stroker = 100 LC	

MERCURY	$74.40
1954 (std. bore 3-5/8)	3-3/4

OLDSMOBILE	$87.20
1957-58	4
Flat Top	4-1/8

OLDSMOBILE	$92.00
1957-58 Domed	4
High comp. approx. 11.5	4-1/8
plus increase from bore	

OLDSMOBILE	$77.60
1949-56	3-7/8
	3-15/16
	4

OLDSMOBILE	$82.40
1949-56 Domed Hi comp.	3-7/8
9.5:1 comp. ratio with	3-15/16
1949-55 Heads	4
10.5:1 comp. ratio with	
1956 Heads	

OLDSMOBILE	$82.40
1949-56	3-7/8
1/8 low comp. for	3-15/16
1/4 stroker	4

PACKARD	$104.00
1955-56 Std. bore	4-1/8
Patrician & Carriben	4-1/4
4-1/8 in 1956	4-5/16

PACKARD	$104.00
1955	3-13/16
10.5:1 (5540) comp. ratio	

PLYMOUTH	$96.80
1956	3-3/4
10.0:1 comp. ratio	3-7/8
	3-15/16

PLYMOUTH V8	$84.00
1955	3-9/16
	3-5/8

PLYMOUTH V8	$96.80
1955	3-9/16
10.0:1 comp. ratio	3-5/8

PONTIAC	$94.40
1957	3-15/16
	4

PONTIAC V8	$77.60
1955-56	3-3/4
	3-13/16
	3-7/8
	3-15/16
	4

PONTIAC V8	$82.40
1955-56	3-3/4
10.0:1 comp. ratio	3-13/16
	3-7/8
	3-15/16
	4

STUDEBAKER V8	$77.60
1951-58	3-9/16
3-5/8 stroke eng.	3-5/8

STUDEBAKER V8	$77.60
1951-58	3-9/16
For 3-1/4 stroke eng.	3-5/8

SPEED ENGINE WARM UP and HEATER ACTION ON COLD DAYS!

Eliminate Radiator Fan
and Add 8 to 15 Horsepower with

EVERKOOL POWER BOOSTER KIT

Eliminate Fan! The radiator fan was designed to keep the engine cool, but it operates whenever the engine is running—on cold days as well as hot days. And when the engine is started on a cold day, the fan actually delays the desired engine warm-up and heater action!

That fan is really necessary for cooling at low speeds when the engine is hot, like in traffic jams or on hills. At speeds faster than 30 MPH, the fan just gets in the way of cooling air which normally flows through the radiator and easily provides enough cooling action to do the job. Here, the fan is actually robbing the engine of 8 to 15 h.p. every second!

FITS ALL CARS AND LIGHT TRUCKS

Ideal for every car owner. Fits all cars, delivery trucks, sports cars, racing cars—one kit universal for 6- or 12-volt motors. Effective with both High and Low Temperature Thermostats.

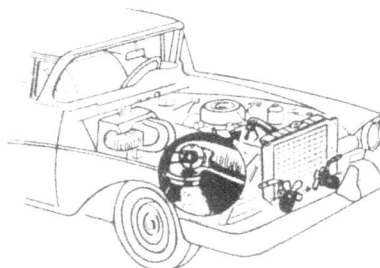

The heart of the EVERKOOL POWER BOOSTER KIT is a pair of fans, two low voltage automotive motors, plus Temp-trol—EVERKOOL's exclusive automatic temperature control.

The fans are easily installed on opposite sides of the radiator near the bottom, so as not to block the air rushing in through the radiator. When the liquid in the cooling system heats up to 185°, the unit kicks on. With both fans working independently of the engine, the radiator cools down fast. When the temperature is restored to normal, the entire mechanism shuts off.

E-300 Kit – Thermostatic or manual
control - - #3866   List $38.75

E-200 Kit – Manual Control
        # 3889   List $32.75

Easy to Install The No. E300 EVERKOOL POWER BOOSTER KIT comes complete with all parts necessary for easy installation, including: motors, fans, automatic temperature switch, toggle switches, and assorted nuts, bolts, brackets, lock washers, plus easy-to-follow illustrated instructions.

SAFETY HUBS
Required by many racing organizations and drag strips. Designed primarily for Ford Rear Ends, they may be adapted to other makes with simple alterations. The assembly is installed between brake drum and axle housing to prevent loss of wheel in the event of a broken axle. Kit includes two safety hubs, two rings, and instructions.
No. 3473 . . . . Pair $20.95

FAN-O-MATIC

MODEL 200
Automatic FAN RELEASE

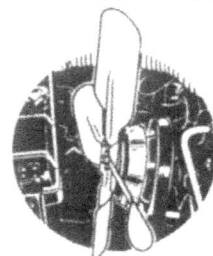

At speeds over 35 MPH, standard automobile fans waste horsepower! The Fan-O-Matic now makes your fan a free-wheeling unit, releasing wasted horsepower at speeds over 40 MPH.
• Automatically disengages fan at speeds over 40 MPH
• Increases Available Horsepower ...up to 15 h.p.
• Increases Gas Mileage ... up to 2 mpg
• Increases Acceleration ... up to 10%
• Reduces Engine Noise
• Reduces Vibration
Tested and proved at Indianapolis Speedway under supervision of United States Automobile Club. Simple to install, adaptable to all cars.
Specify year, make and model.
NO. 3881 Fan-O-Matic $39.50

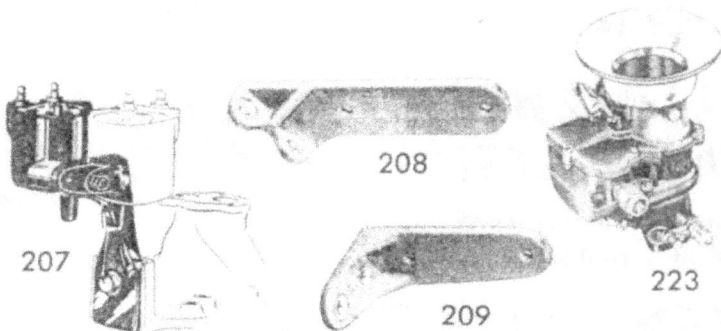

FLARED CARBURETOR STACK

223—Venturi type stacks to improve air flow through the carburetors. Polished aluminum... ¼ 2.00

DUAL COIL BRACKETS

207—Sturdy design, cast aluminum. Made for two 42-48 Ford coils. 42-48 Ford and Mercury fits in place of fan bracket on manifold....... ½ 3.75

208—49-56 Ford and Mercury fits in stock position on the head................... ¼ 2.75

209—Universal firewall mounting type....... ½ 2.75

DEGREE WHEEL

222—For setting and checking valve or ignition timing. Cast aluminum, polished finish, 8" diameter. Divided in degrees. Fitted with set-screw. 1⁵⁄₁₆" hub hole 1 6.95

QUAD AIR CLEANER

225—For low hoods. 9½" diameter, 2½" high air cleaner for Stromberg, Carter and Rochester quad carburetors. Polished aluminum........ 3½ 11.75

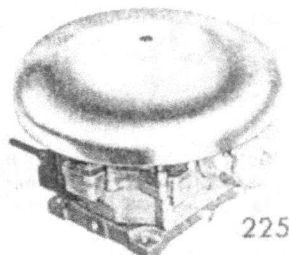

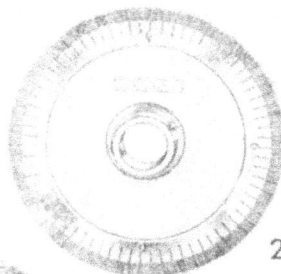

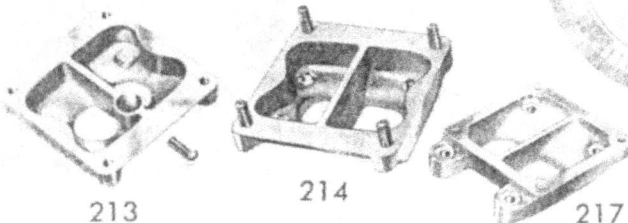

SPECIAL AIR CLEANERS

CARBURETOR ADAPTERS	Shpg. Wt. Lbs.	List Price
213—Ford V8 type 3-bolt dual throat manifold to quad carburetor	1	$6.75
214—Standard 4-bolt dual throat manifold to quad carburetor	1	6.75
217—Chrysler and De Soto type 4-bolt dual throat to quad carburetor...................	1	7.75

These are special air cleaners made especially for various special manifold applications.

HELLINGS AIR CLEANERS

S503 for Holley Quad-Jet Carburetor—Mesh type—8½" O.D.
Price $ 9.67   Cat. #3580

S420 for Carter Quad-Jet Carburetor—Mesh type—8½" O.D.
Price $ 9.67   Cat. #3581

D503 for Holley Quad-Jet Carburetor—Bonnet type—8¼" O.D.
Price $12.95   Cat. #3582

D420 for Carter Quad-Jet Carburetor—Bonnet type—8¼" O.D.
Price $12.95   Cat. #3583

SD420 for Carter Quad Carburetor——Mesh type   7" O.D.
Price $8.95   Cat # 3890

SD503 for Holly Quad Carburetor——Mesh type   7" O.D.
Price $8.95   Cat # 3891

H-C AIR CLEANERS

4-700 for Carter Quad-Jet Carburetor—Bonnet type—11¼" O.D.
Price $14.95   Cat. # 3584

4-723 for Carter Quad-Jet Carburetor—Bonnet Oil bath—11¼" O.D.
Price $21.95   Cat. # 3585

4-702 for Holley Quad-Jet Carburetor—Bonnet type—11¼" O.D.
Price $14.95   Cat. # 3586

4-724 for Holley Quad-Jet Carburetor—Bonnet Oil bath—11¼" O.D.
Price $21.95   Cat. # 3587

4-711 for Stromberg 97's, 48's and Ford 3-bolt Carburetors—4½" O.D.
Price $ 4.25   Cat. # 3588

Special Note ... All air cleaners listed for Carter Quad-Jets will also fit Stromberg and Rochester Quad-Jet carburetors.

4-7S for Thunderbird Triple-Dual manifold—4½" low
$ 5.50   Cat. # 3589

4-24S for Mercury and Thunderbird dual quad-jet Carter carb.
$14.95   Cat. # 3590

C265 for Chevrolet V8—5½"
$ 7.50   Cat. # 3591

R2516 for Chevrolet 6 dual manifold—7"
$ 7.50   Cat. # 3592

NEW IMPROVED SAFETY BELTS ...

Made by Ray Brown of Hollywood.

Nylon webbing, chromium buckle, complete with all fittings. Colors: tan, white, black, blue, green, gray, or maroon. Specify.

$9.95 ea.   Cat. #1028

FILT-O-REG

Combination Fuel Pressure Regulator and Filter. Provides a constant, even fuel pressure of only two pounds to the carburetor float valve and seat, maintaining the PROPER float level under ALL driving conditions. Cuts gas consumption by reducing fuel waste. Prevents stalling due to carburetor flooding. No adjustments are necessary. Keeps dirt and water out. Eliminates vapor-lock. Easy to install—fits in the fuel line between fuel pump and carburetor. Guaranteed to exert a maximum 2 lb. constant, even pressure at the carburetor under all driving conditions.

Regular Model FILT-O-REG .... # 3578

WE DISTRIBUTE ALL LEADING BRANDS OF CAMS LITERATURE PERTAINING TO SAME FURNISHED ON REQUEST.

THOMAS MAGNESIUM ROCKER ARMS NOW AVAILABLE WITH
BRONZE OILITE BUSHINGS.
Special rocker arms with bronze oilite bushings and longer adjustment screws. To use these special arms it is necessary to use
special metal valve cover gaskets. #880 for Cadillac and #882
for Oldsmobile.

#700	Cadillac	'49 Std. Ratio	$45.75	Cat. #3905
#701	Cadillac	'49 Highlift Ratio	45.75	Cat. #3906
#702	Cadillac	'50-'54 Std Radio	45.75	Cat. #3907
#703	Cadillac	'40-'58 Hi Lift Ratio	45.75	Cat. #3908
#704	Oldsmobile	'49 Std Ratio	46.50	Cat. #3909
#705	Oldsmobile	'49 Hi Lift Ratio	46.50	Cat. #3910
#706	Oldsmobile	'50-'51 Std Radio	45.75	Cat. #3911
#707	Oldsmobile	'50-'51 Hi Lift Ratio	45.75	Cat. #3912
#708	Oldsmobile	'52-'56	45.75	Cat. #3913
#709	Oldsmobile	'57-'58	45.75	Cat. #3914

Metal Valve Cover Gaskets available for Cadillac, Chevrolet & Oldsmobile

#800	Cadillac	49-58	$ 7.45	Cat. #3915
#881	Chevrolet	55-58	6.75	Cat. #3916
#882	Oldsmobile	49-57	7.25	Cat. #3917

THOMAS TUBULAR PUSH RODS ... eliminates valve train
fatigue and helps eliminate excessive cam wear. Made of
strong, light weight tubing and finished with black dulite.
They will not rust.
STEEL PUSH RODS
Model

#1060	Cadillac '49-'51	$24.00	Cat. #3190
#1061	Cadillac '52-'56	24.00	Cat. #3191
#1062	Ford Six Cyl. '52-'56	21.00	Cat. #3192
#1063	Ford V8 '54-'56	24.00	Cat. #3193
#1064	Mercury V8 '54-'56	24.00	Cat. #3194
#1065	Oldsmobile '49-'51	24.00	Cat. #3195
#1066	Oldsmobile '52-'56	24.00	Cat. #3196

THOMAS HYDRAULIC TAPPET CONVERSION KITS

Designed to convert your hydraulic tappets into manually
adjusted tappets. Simple in design, easily installed. Note
many kits include tubular push rods.
Model

#1033	Buick Dynaflow Straight 8 '49-'53	$11.70	Cat. #3179
#1034	Buick V8 '53-'56 (16 tubular push rods)	39.50	Cat. #3180
#1050	Buick V8 w/push rods '57-'58	39.50	Cat. #3899
#1035	Chevrolet Power Glide 235 '51-'58	10.80	Cat. #3181
#1051	Chevrolet V8 '55-'58	16.00	Cat. #3900
#1036	Cadillac, Flathead '41-'47, early '48	19.75	Cat. #3182
#1045	Cadillac Flathead, Late 1948	19.75	Cat. #3299
#1046	Cadillac '49-'58 w/push rods	39.50	Cat. #3301
#1037	Chrysler '51-'52 early 1953	19.75	Cat. #3183
#1038	Chrysler late 1953-56 (16 tubular push rods)	39.50	Cat. #3184
#1039	DeSoto 1952	19.75	Cat. #3185
#1040	DeSoto '53-'56 (16 tubular push rods)	39.50	Cat. #3186
#1042	Dodge '53-'56 (16 tubular push rods)	39.50	Cat. #3187
#1043	Lincoln Flathead '49-'51	19.75	Cat. #3188
#1044	Lincoln '52-'56	10.70	Cat. #3189
#1047	Oldsmobile '49-'51 w/push rods	39.50	Cat. #3302
#1048	Oldsmobile '52-'56 w/push rods	39.50	Cat. #3303
#1049	Oldsmobile '57-'58 w/push rods	39.50	Cat. #3901
#1052	Pontiac V8 '55	16.00	Cat. #3902
#1053	Pontiac V8 '56	16.00	Cat. #3903
#1054	Pontiac '57-'58	16.00	Cat. #3904

THOMAS MAGNESIUM ROCKER ARMS

ADJUSTABLE - - eliminates hydraulic tappets. By using high
lift rocker arms you increase the effective valve opening
without changing the camshaft. Lighter weight means more
R.P.M.'s without floating. Uses stock push rods.

# 999	Chevrolet '55-'56 V8 Assembly, Std. Lift	$94.75	Cat. #3411
#1000	Chevrolet '55-'56 V8 Assembly Hi Lift	94.75	Cat. #3315
#1001	Cadillac '49 Standard	39.50	Cat. #3149
#1002	Cadillac '49 Hi Lift	39.50	Cat. #3150
#1005	Cadillac '50-'54 Standard	39.50	Cat. #3151
#1006	Cadillac '50-'56 Hi Lift	39.50	Cat. #3152
#1007	Chevrolet 216 '41-'53 Hi Lift	32.50	Cat. #3153
#1009	Chevrolet 216 '41-'53 (intake only)	16.50	Cat. #3304
#1008	Chevrolet 235 '51-'55 Hi Lift	32.50	Cat. #3154
#1010	Chevrolet Corvette '54-'55 Hi Lift	32.50	Cat. #3306
#1011	Ford Six Cyl. '52-'56 Standard	32.50	Cat. #3307
#1012	Ford Six Cyl. '52-'56 Hi Lift	32.50	Cat. #3155
#1013	Ford & Mercury V8 '54-'55 Standard	39.50	Cat. #3156
#1014	Ford & Mercury V8 '54-'56 Hi Lift	39.50	Cat. #3157
#1015	GMC 270 Hi Lift	32.50	Cat. #3158
#1016	'54-'55 Mercury Standard	39.50	Cat. #3893
#1017	'54-'57 Mercury Hi Lift	39.50	Cat. #3894
#1018	Oldsmobile '49 Standard	41.25	Cat. #3161
#1019	Oldsmobile '49 Hi Lift	41.25	Cat. #3162
#1020	Oldsmobile '50-'51 Standard	39.50	Cat. #3163
#1021	Oldsmobile '50-'51 Hi Lift	39.50	Cat. #3164
#1022	Oldsmobile '52-'56	39.50	Cat. #3165
#1023	Studebaker '51-'54 Standard	39.50	Cat. #3166
#1026	Oldsmobile '57-'58	39.50	Cat. #3895
#1027	Ford '58 Hi Lift	39.50	Cat. #3896
#1028	Mercury '58 Hi Lift	39.50	Cat. #3897

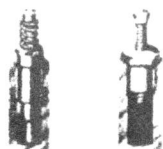

COMPLETE ROCKER ARM ASSEMBLIES:
New Lifters, push rods, rocker arms, rocker shafts and
aluminum spacers for valve cover clearance - Complete.

CADILLAC

Standard Lift	1949-56	Cat. #3479	$151.25
High Lift	1949-56	Cat. #3480	$151.25

OLDSMOBILE

Standard Lift	1949-51	Cat. #3478	$144.50
High Lift	1949-51	Cat. #3477	$144.50
	1952-58	Cat. #3898	$144.50

Number			
3882	Ford - Chrysler - Pontiac 1/2" Hex 5/16-24 Thread Male, ea.	$	1.00
3883	Chevy Corvette only, ea.		1.00
3884	Cadillac - Chevrolet Stock Valve Covers, ea.		1.00
3885	Cadillac - Chevrolet Custom Finned Covers Only, ea.		1.00
3886	Chrysler Only (5/16-24 Internal Thread) Use with Stock Studs, ea.		1.00

WING NUTS

THESE BEAUTIFULLY CHROMED ELEVATED WING NUTS PROVIDE THE ULTIMATE IN VALVE
COVER NUTS. THE SHANK OF THE NUT IS CENTERLESS GROUND BEFORE PLATING IN ORDER
THAT THE FINEST POSSIBLE PLATING MAY BE ACCOMPLISHED. THE LARGE WING PERMITS
REMOVAL OF THE VALVE COVERS WITHOUT THE USE OF ANY TOOLS.

Number		List		Number		List
3626	Chev V-8; Cad Stock	$ 2.00		3628	Custom Chev-8 (Finned)	$ 2.00
3627	Pontiac; GMC Stock	2.00		3629	Chev Corvette Only	2.00

BREATHER
POLISHED ALUMINUM

Designed to Relieve
Pressure in Valve Covers

Gives Your Engine that
Professional Appearance

Very Popular - Attractive

Mounts on Most Stock
Valve Covers. Small -
easy to install

If breathers are to be installed on custom style valve
covers, it will be necessary to remove a fin.

No. 3661 Breather .... $ 4.95 ea.

WING NUTS
FOR VALVE COVERS

Highly chromed -- very popular -- adds
that final touch of distinction to your
custom engine.

EACH

NO. 3509 Tapped 5/16 -24 thread $1.00
NO. 3510 Tapped 3/8 -24 thread $1.00

UNIVERSAL STUDS:

For valve covers long enough for most installations —
just cut for your length.

#3511   3/8" std thread one end S.A.E. other....$ .50

#3512   5/16" std thread one end S.A.E. other.... .50

VAC-U-TRIM
by OMEC
Patent pending

ONLY
$5.95

THE ONLY VACUUM ADVANCE MICROMETER ADJUSTING UNIT MANUFACTURED TODAY

The precision micrometer-like adjustment in Vac-U-Trim offsets the lack of vacuum control resulting from multiple carburetion

- Ends ignition adjustment troubles on late or custom type engines.
- Enables you to make micrometer-like adjustment on the ignition of your car.
- Provides maximum engine performance, eliminates engine ping without loss of horse power.
- Enables adjustment of ignition to atmospheric conditions and changes in altitude
- Eliminates engine lag caused by sudden acceleration.
- Reduces stalling due to unbalanced distributor action.
- Improves gasoline mileage.
- Maximum engine performance through precision ignition control.

Order No. for all cars   #3666
Except Ford, Mercury and Lincoln '49 through early '56   #3667

MAGNETO OMEC-DISTRIBUTOR POSITION INDICATOR

This Aluminum Degree Plate has a patented gum back which can be applied to any Magneto or Distributor. You merely strip the protective paper from the gummed back - place the indicator in position and press firmly. The Chrome Plated Indicator Arm is then attached to the Engine or Firewall. Accurate timing changes are then possible as well as re-positioning to a previous setting.

3892   Omec Mag-Distributor Indicator   - list $ 1.90

CHEVY FLYWHEEL SAFETY BOLTS By OMEC

OMEC DESIGNED THESE PRECISION GROUND CHROME MOLY BOLTS TO REDUCE THE HAZARD OF BLOWN CLUTCHES IN COMPETITION CHEVY ENGINES. THEY ARE PROPERLY HEAT-TREATED FOR MAXIMUM STRENGTH AND CENTERLESS GROUND TO WITHIN .001 FOR PERFECT FIT. EACH BOLT IS INDIVIDUALLY IN-SPECTED AND TAGGED BEFORE PACKAGING. IT IS RECOMMENDED THAT THE DOWEL PIN BE REMOVED WHEN USING OMEC FLYWHEEL SAFETY BOLTS. THESE BOLTS ARE APPROXIMATELY .034 LARGER THAN STOCK LOW CARBON FLY-WHEEL BOLTS AND ALLOW ABOUT .004 CLEARANCE IN THE PRESENT FLYWHEEL BOLT HOLE.
THEY WILL ELIMINATE THE FLYWHEEL WHIP EXPERIENCED WITH STOCK F-W BOLTS. CUSTOM MADE F-W's MAY BE FITTED TO THE BOLTS FOR MAXIMUM SAFETY. ADAPTABLE TO ALL CRANKS TAPPED 7/16-20 THREAD.

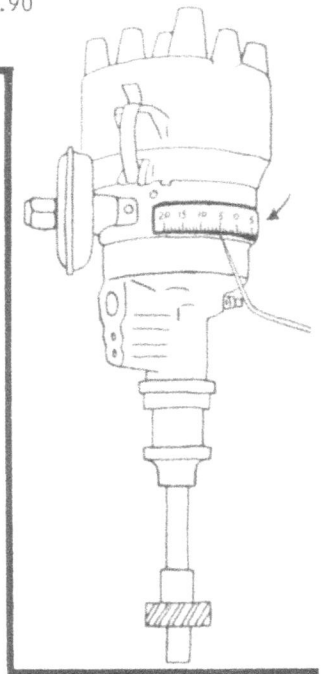

"FOR SAFETY SAKE"

Chev. Flywheel Safety Bolts -
#3669   set of 6 $ 9.95

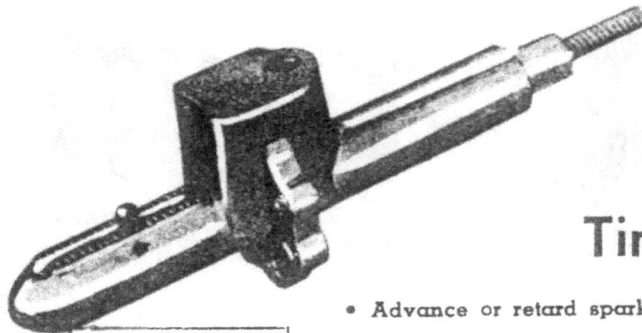

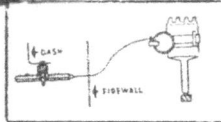

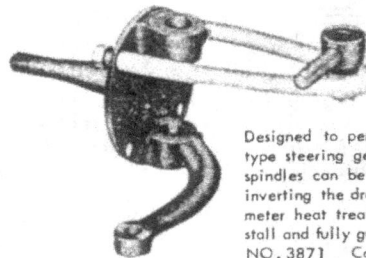

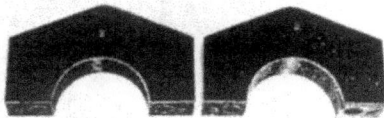

AT LAST!

SAFETY HELMETS

by PELA

These famous helmets are used all over the world! Weight approximately 20 ounces. Shells are white trimmed in rich brown. (Repaint or decorate if you desire.)

INSIDE CONSTRUCTION IS IMPORTANT!

Four strips of strong webbing lie beneath the flaps drawn together by a cord which permits a certain amount of adjustment.

NUMBER "THIRTY-TWO" – "THE ONE FOR THE ROAD"

Aluminum shell with double nape guard, plastic fittings of imported quality. A lot of rugged protection for so little money. Widely used by club riders, in scramble events, for boat racing, etc.

BARGAIN PRICED AT ONLY **$14.95**   Part #3610

ALWAYS ORDER HELMETS BY HAT SIZE

NUMBER "THIRTY-SIX"

The "PERFECT THIRTY-SIX" — Aluminum shell with extraordinary strong and padded nape guard, top quality GENUINE LEATHER LINING. Thoroughly a luxury helmet. The neck flap can be buttoned up, which is a very popular feature of PELA HELMETS — Keeps dirt from going down the wearer's back.

PRICE **$19.95**   Part #3611

ALWAYS ORDER HELMETS BY HAT SIZE

NUMBER "THREE-SIXTY"

Professionally accepted, APPROVED BY NASCAR. Best all-around same luxury finish as the "PERFECT THIRTY-SIX" — BUT, it has a PERLON SHELL. PERLON is a revolutionary German material. "Soft as Rubber" — "Strong as Steel"!! A very strong brim that is reinforced. Shell will withstand most severe abuse and is non-inflammable, non-absorbent.

PRICE **$29.50**   Part #3612

ALWAYS ORDER HELMETS BY HAT SIZE

TIE CLASP: Intended for exclusive wear by men and women (amateurs as well as professionals) who engage in a sport in which a safety helmet is worn, such as Bob Sledding, Ski Racing, Speed Boating, Motorcycling, Auto Racing of all categories. None of these sports is for "sissies"; so identify yourself with a red-blooded sport! WEAR A HELMET! **$1.50** individual gift box packed

Part #3613

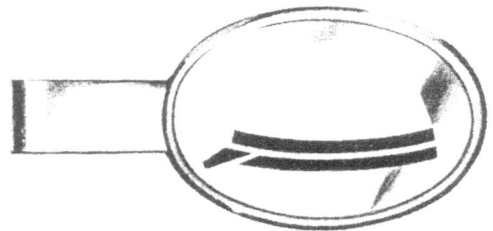

HARD FACED OVERLAY KITS

By *Clay Smith Engineering*

MR. HORSEPOWER

FRUSTRATOR - - - - - - - Dual purpose grind for city traffic and week-end dragging. Excellent short track grind with proper equipment.

MATCHLESS - - - - - - - Designed for maximum Horse Power recommended specifically for — "ALL OUT" Competition Engines. Not for Automatic Transmissions.

PEERLESS - - - - - - - - Sacrifices low speed Torque to develop EXTREME R.P.M. — Recommended ONLY for BONNEVILLE or MARINE USE.

FORD — MERCURY — THUNDERBIRD

KIT ONLY $85.00   **CAMSHAFT ONLY $135.00**   **COMPLETE $220.00**

Ready to install — NO REWORK — Kit consists of 16 Lightweight tubular push rods. 16 Chilled iron lifters. 16 Special Outer Valve Springs and Dampners. 16 Heavy duty spring retainers — Heat Treated — Not Case Hardened.

Grind	Timing	Lift	Clearance
Frustration 272	In. 24-68 Ex. 68-24	.415	.016 .016
Matchless 284	In. 30-74 Ex. 74-30	.423	.016 .016
Peerless 296	In. 38-78 Ex. 78-38	.412	.016 .016
BBX	In. 20-72 Ex. 72-20		

For Superchargers only. Idles quiet very effective. H.P. output throughout. All stages of R.P.M.

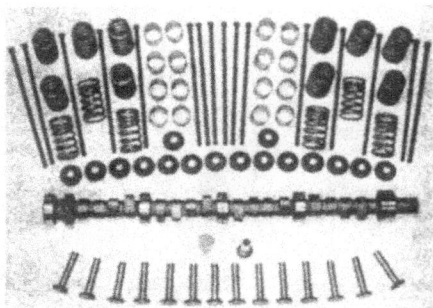

CHEVROLET V-8 — AVAILABLE IN ALL ABOVE LISTED GRINDS

"ALIZED" — same price — $135.00 Cam Only

TWO KITS AVAILABLE

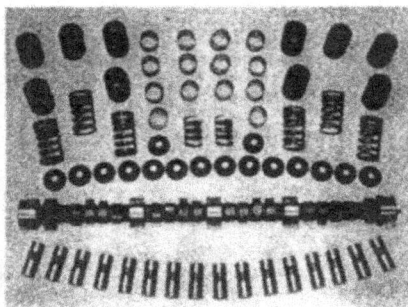

KIT A   **PRICE $50.00**
For Alized or Processed Re-Grinds consists of following: 16 chilled iron lifters—OIL CONTROLLED. 16 heavy duty dampner type springs—designed to not necessitate machining of Cylinder Heads — to be used with standard push rods and standard retainers—NO REWORK NECESSARY.

✳ **KIT B**   **PRICE $74.00**
ALL OUT RACING—All 272 - 284 - 296 grinds. Consists of following: 16 Chilled iron lifters. 16 Large diameter dampner type valve springs. 16 Special retainers—HEAT TREATED—Use stock rocker arms & push rods.

✳ NOTE: Special Cutter for enlarging spring nest in cylinder head necessary—$20.00 deposit required, refundable if returned within 30 days.

CHEVROLET V8 — Processed Re-Grinds. Available in only 272 — "Frustrator" grind. Price including core **$86.40**

NOTE: Kit A should be used for best results with this camshaft.

FOR PROFESSIONAL USE — Steel billet camshaft available.

ALL OTHER GRINDS — A complete listing of Clay Smith camshafts is being prepared therefore we are in a position to furnish your needs on this nationally accepted merchandise.

"Alized" grinds available for all other O.H.V. engines — specify $135.00

KITS FOR ABOVE CAMS

PONTIAC
Kit A — No rework necessary $49.00
*Kit B — Special $74.00

OLDSMOBILE
Kit A — Complete Allout $126.00
*Kit B — $100.00
*49 - 51 Special cutter

CADILLAC
*Kit A — Complete Allout $126.00
Kit B — $100.00

CHRYSLER, DESOTO, DODGE, PLYMOUTH
*Kit — Complete Allout $103.00

*NOTE requires cutter for enlarging spring nest in cylinder head - $20.00 deposit refundable.

DOUGLASS CHROME SPECIALTIES

LAKES PLUGS, SIDE TAIL PIPES, AND STACKS ARE BOXED IN PAIRS. THEY MAY BE ORDERED INDIVIDUALY. SPECIFY HALF PAIRS

CAT. NO.	OVERALL LENGTH	DIMENSIONS BETWEEN BENDS	PIPE SIZE	PART NO.	LIST PER PAIR	CAT NO.	KIT NO.	LIST
3706	14"	Straight	1¾"	LP 134-1	$13.05	3717	KL-131	$ 9.50
3707	14"	Straight	2"	LP 200-1	13.05	3718	KL-132	10.50
3708	14"	Straight	2"	LP 200-2	16.40	3719	KL-132	10.50
3709	34½"	22"	1¾"	LP 134-2	28.90	3720	KL-133	13.90
3710	40"	26½"	1¾"	LP 134-3	28.90	3721	KL-133	13.90
3711	40"	26½"	2"	LP 200-3	28.90	3722	KL-134	14.90
3712	34½"	22"	2"	LP 200-4	28.90	3723	KL-134	14.90
3713	76"	60"	1¾"	LP 134-6	60.00	3724	KL-135	15.60
3714	76"	60"	2"	LP 200-5	60.00	3725	KL-136	16.60
3715	26½"	15"	1¾"	LP 134-1A	18.90	3726	KL-133	13.90
3716	26½"	15"	2"	LP 200-1A	18.90	3727	KL-134	14.90

New Douglass triple chrome plated Lakes Plugs give your car that all New, Glittering and Shiny, Sporty Custom Look. Pipes retain their lustrous high polish, even in the worst weather conditions. Two superbly engineered sparkling chrome bends create the beautiful Stylish Chrome Rail running along the lower edge of the body. New glittering Chrome Custom Finned Blocking Plates and Round Head Bolts add the finishing touch to the created beauty of all Douglass Chrome Lakes Plugs. These Douglass Lakes Plugs are made from one piece of tubing (not universal). They are custom engineered and bent to fit any car and give you the maximum added performance possible. Available from 22" to 60" between bends. Easy to order. Simply select pipe length and size you desire and order by part number. For your convenience special Douglass Lakes Plug Kits are available. These kits will simplify the installation (only one weld needed) and have all the parts included to give your car maximum beauty and performance. For perfection, order your Douglass Kit.

3728	26½"	15"	1¾"	STP 134-1	14.80	3735	KS-141	12.90
3729	26½"	15"	2"	STP 200-1	14.80	3736	KS-142	14.90
3730	34½"	22"	1¾"	STP 134-2	15.50	3737	KS-141	12.90
3731	40"	26½"	1¾"	STP 134-3	17.60	3738	KS-141	12.90
3732	40"	26½"	2"	STP 200-6	17.60	3739	KS-142	14.90
3733	76"	60"	1¾"	STP 134-4	52.00	3740	KS-143	7.80
3734	76"	60"	2"	STP 200-5	52.00	3741	KS-144	8.80

Douglass triple chrome plated Side Tail Pipes add lasting glittering beauty to your car. These beautiful Side Tail Pipes were especially engineered and designed to help eliminate power destroying back pressure created by exhaust gases. Douglass Side Tail Pipes are easily installed. They connect to the rear of your muffler. The exhaust gases are quickly carried to the outside of your car and curved outward just in from the rear wheel. This prevents tail pipe scraping and discoloration of the rear bumper chrome. Douglass chrome (mirror like finish) Side Tail Pipes are all one piece design, custom bent and engineered available to fit any car Select your desired size and length. Immediate delivery on all sizes from 15" to 66" between bends. 66" Side Tail Pipes have special Y connections for easy connection to your muffler. We suggest the ready-made Douglass installation kit to complete your Side Tail Pipe installation.

3742	66"	60"	1¾"	VS134	42.78	3751	KT-151	15.60
3743	66"	60"	2"	VS200	42.78	3752	KT-152	17.90
3744	68"	60"	2"	VS201 R & L	44.45	3753	KT-154	11.20
3745	66"	. .	1¾"	VS134-1	34.45	3754	KT-151	15.60
3746	66"	. .	2"	VS200-1	34.45	3755	KT-152	17.90
3747	66"	. .	1¾	VS134-2	42.78	3756	KT-151	15.60
3748	66"	. .	2"	VS200-2	42.78	3757	KT-152	17.90
3749	78"	Ford Rail	2"	HS200-R and L	73.05	3758	KT-156	21.00
3750	78"	Chev Rail	2"	HS201-R and L	73.05	3759	KT-156	21.00

Douglass Triple Plated Chrome Pick-up Stacks are fully guaranteed to have the highest quality, heavy duty chrome plating available . . . shine and glitter even under adverse conditions. Smartly engineered and tailored to fit all pick-ups. Available in four custom designs in the vertical types; either 1¾" or 2" pipe dia. size. The Douglas Deluxe Horizontal Rail type is available for most pick ups . . . 2" pipe dia. size only. Order by part number or for complete easy installation specify the Douglas Beauty and Performance Kit number.

The above Side Tail Pipes, Lake Plugs and Truck Stacks are custom bent for ease of installation. Plated copper, nickel and chrome. There are no additional chrome bends needed. The installation kits recommended for the above chrome accessories contain the correct number of chrome clamps, 90° bends, slip connections and plain clamps (for under the chassis). Complete easy to follow installation instructions are included with every pair of the above chrome pipes. Welding equipment is essential for installing Lake Plugs. Side Tail Pipes and Truck Stacks may be installed with ordinary hand tools.

FINNED BLOCKING PLATES

The New Triple Chrome Plated Raised Finned Design. The very latest for complimentary beauty. Absolutely a must for a real smart custom look. Furnished with plated Round Head Bolts for a streamlined effect. This item is included in all Douglass Lake Plug Kits.

CAT. NO.	TUBE & FLANGE SIZE	STANDARD PACKAGE	LIST EACH	PART NO.
3760	1¾"	6	$3.34	BP63-C
3761	1¾"	6	4.18	BP63-CX
3762	2	6	3.34	BP62-C
3763	2	6	4.18	BP62-CX

FLAT STEEL TYPE CHROME

CAT. NO.	TUBE & FLANGE SIZE	STANDARD PACKAGE	LIST EACH	PART NO.
3764	1¾"	6	$2.10	BP3-C
3765	1¾"	6	2.84	BP3-CX
3766	2	6	2.10	BP2-C
3767	2	6	2.84	BP 2 CX
3768	2 (3 hole)	6	3.00	BP5-C
3769	2 (3 hole)	6	4.11	BP5-CX

CX NUMBERS — CHROME NUTS, BOLTS & WASHERS INCLUDED

LAKES PLUGS, SIDE TAIL PIPES AND STACKS ARE BOXED IN PAIRS. THEY MAY BE ORDERED INDIVIDUALY. SPECIFY HALF PAIRS.

CHROME AND PLAIN CLAMPS AND BRACKETS

CAT. NO.	SIZE	FINISH	DESCRIPTION	STANDARD PACKAGE	LIST EACH	PART NO.
3770		Plain	Angle Brkt.	6	$ .30	L1
3771		Chrome	Angle Brkt.	6	.90	L1-C
3772		Plain	Angle Brkt.	6	.30	L2
3773		Chrome	Angle Brkt.	6	.90	L2-C
3774	1¾"	Plain	Clamp with Brkt.	6	1.05	358-A
3775	1¾"	Chrome	Clamp with Brkt.	6	3.05	358-AC
3776	2	Plain	Clamp with Brkt.	6	1.05	36
3777	2	Chrome	Clamp with Brkt.	6	3.05	36-C

CHROMED BOLT WITH PLATED NUT AND WASHER

CAT. NO.	SIZE	DESCRIPTION	STANDARD PACKAGE	LIST EACH	PART NO.
3778	5/16-18 x 1½	Carriage Bolt	12	$ .42	CB516-C
3779	5/16-18 x 1½	Cap Screw	12	.42	CS516-C
3780	7/16-14 x 1	Cap Screw	12	.42	CS716-C
3781	3/8-14 x 1	Carriage Bolt	12	.42	CB380-C

90% BEND AND S BEND PLAIN NOT CHROMED

CAT. NO.	PART NO.	SIZE	RADIUS	DEGREE OF BEND	LIST
3782	B134	1¾"	6"	90°	$2.00
3783	B134-S	1¾"	6"	90° Expanded one end	2.50
3784	B200	2"	6"	90°	2.50
3785	B200-S	2"	6"	90° Expanded one end	3.00
3786	SB134	1¾"	6"	Offset	3.00
3787	SB134-S	1¾"	6"	Offset Expanded one end	3.50
3788	SB200	2"	6"	Offset	3.50
3789	SB200-S	2"	6"	Offset Expanded one end	4.00

specify DOUGLASS for the BEST
ENGINEERING — WORKMANSHIP — CONSTRUCTION

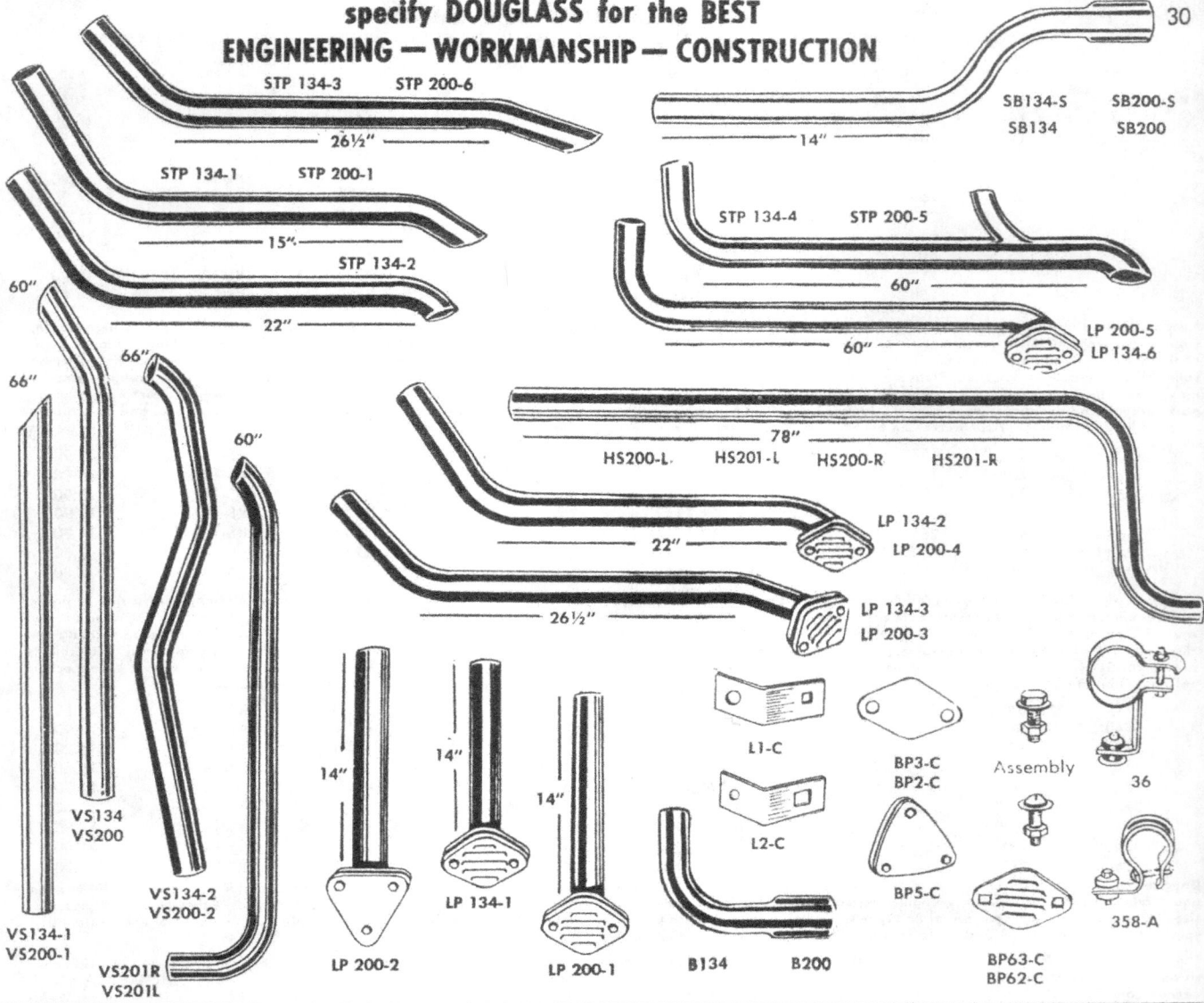

STP 134-3   STP 200-6

26½"

SB134-S   SB200-S
SB134   SB200

STP 134-1   STP 200-1

15"

STP 134-2

STP 134-4   STP 200-5

22"

60"

60"

60"

60"

LP 200-5
LP 134-6

66"

66"

78"

60"

HS200-L   HS201-L   HS200-R   HS201-R

66"

60"

22"

LP 134-2
LP 200-4

26½"

LP 134-3
LP 200-3

VS134
VS200

VS134-2
VS200-2

14"

14"

14"

L1-C

BP3-C
BP2-C

Assembly

36

VS134-1
VS200-1

VS201R
VS201L

LP 200-2

LP 134-1

LP 200-1

L2-C

B134   B200

BP5-C

358-A

BP63-C
BP62-C

Lakes Plugs

LP134-1
LP200-1
LP200-2

LP134-2
LP200-4

LP134-3
LP200-3

LP134-6
LP200-5

Side Tail Pipes

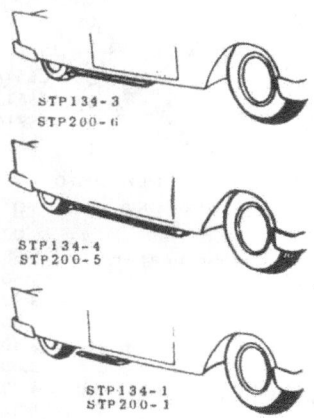

STP134-3
STP200-6

STP134-4
STP200-5

STP134-1
STP200-1

Vertical & Horizontal Stacks

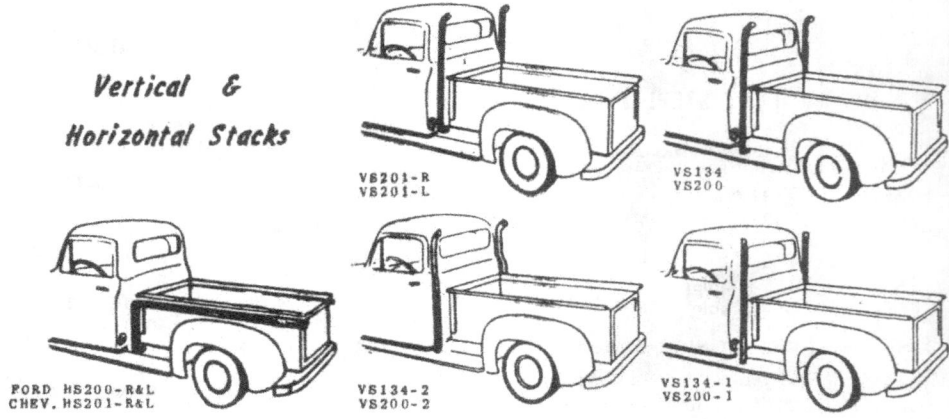

VS201-R
VS201-L

VS134
VS200

FORD HS200-R&L
CHEV. HS201-R&L

VS134-2
VS200-2

VS134-1
VS200-1

STELLING & HELLINGS BREATHERS

Just as today's high winding, large displacement engines build terrific pressures in their combustion chambers, pressures are generated in the crankcase on the downward stroke, which collect, reducing efficiency and forcing out the supply of vital oil. Factory designed breathers are inadequate for hopped-up engines, thus you will find these cast aluminum units a must for top performance. Available in a variety of shapes and sizes of $5.95 each.

Cat. No. 3474

AXLE KEYS

Manufactured from Aircraft quality Chrome-Alloy, these ¼" and ⁵⁄₁₆" square keys are heat treated to Aircraft specifications. Have many times the strength of stock Ford axle keys but will not harm axles and hubs if broken such as tool steel bits do. "Just the right hardness." ¼" keys fit from Model "A" to 1948 Ford and Mercury rear ends. ⁵⁄₁₆" keys fit Model "T" Truck and most all special large axles in racing rear ends. Ideal for OHV, V-8 engine installations in earlier Ford and Mercury chassis. Also pick-up trucks. Used by all the nation's top drag machines.

\# 3632   ¼" Square Keys ....Per Pair $1.50
\# 3633   ⁵⁄₁₆" Square Keys ....Per Pair   2.50

DRAG RACE AXLE KEYS

The Key to a Successful Dragster

S. P. VELOCITY STACK

A major improvement to convert the 97 and 48 Stromberg carburetors on engines where peak performance is required. This top is designed to eliminate drag, friction and turbulence in the airstream, caused by irregularly shaped inlet sections and choke plates in the conventional carburetor. Pressure losses are held to a minimum, resulting in higher volumetric efficiency and horsepower increase. Precision cast of finest aluminum alloy and machined to rigid racing specifications.

Part No. 3471

— SPARK PLUG WIRE —

Plastic transparent wire manufactured to our rigid specifications.

THE DEMAND FOR THIS PRODUCT IS TERRIFIC ALL OVER THE COUNTRY

Realizing this demand we have contracted for this transparent plastic wire, NOT ONLY IN CLEAR, but also in RED TRANSPARENT!!! In other words, we can offer you your choice of either clear or red transparent wire. THESE ARE A MUST. Suggest you ORDER IMMEDIATELY!!!

Free Sample sent upon request.

Red transparent wire	Cat. No. 3622	$15.00 per 100 ft.
Clear transparent wire	Cat. No. 3623	$15.00 per 100 ft.

SAFETY FIRST

WITH A

PRECISION STEEL

FLYWHEEL

SPECIAL FEATURES

- NOT A CASTING
- MACHINED FROM SOLID STOCK
- SPECIAL HEAT-TREATED BOLTS FURNISHED
- FLYWHEEL BOLT HOLES BORED FOR PRECISION FIT TO SPECIAL BOLTS. This way torque is distributed throughout the six bolts instead of the original single dowel pin.
- DIAMETER OF FLYWHEEL FULLY DEGREED
- BALANCED
- ACCURACY HELD THROUGHOUT

FLYWHEELS MADE TO CUSTOMER'S SPECIFICATIONS UPON REQUEST

by PATH

Part #3668   List Price $89.50

Cyl. Diam.	No. Cyls.	No. & Width of Rings Comp.	Oil	Set No.	List	List Chrome Top
CADILLAC, CHRYSLER, OLDSMOBILE						
3 7/8	8	16 – 5/64D	8 – 3/16	2211	$21.00	$26.18
3 15/16	8	16 – 5/64D	8 – 3/16	11581	21.15	27.74
3 15/16	8	8 – 3/32D				
		8 – 3/32	8 – 3/16	3501	21.18	27.49
4"	8	16 – 5/64D	8 – 3/16	2711	20.90	26.80
4"	8	8 – 3/32D				
		8 – 3/32	8 – 3/16	3511	21.18	27.49
4 1/8	8	16 – 5/64D	8 – 3/16	1871	23.20	29.90
CHEVROLET						
3 9/16	6	12 – 3/32	6 – 5/32	3521	16.06	20.79
3 5/8	6	12 – 3/32	6 – 5/32	3531	16.06	20.79
3 3/4	8	16 – 5/64D	8 – 3/16	1191	18.50	24.68
3 3/4	8	8 – 3/32D				
		8 – 3/32	8 – 3/16	3541	21.18	27.49
3 7/8	8	16 – 5/64D	8 – 3/16	2211	21.00	26.18
3 7/8	8	8 – 3/32D				
		8 – 3/32	8 – 3/16	3551	21.18	27.49
3 15/16	8	16 – 5/64D	8 – 3/16	11581	21.15	27.74
4"	8	16 – 5/64D	8 – 3/16	2711	20.90	26.80
CHRYSLER – See Cadillac						
FORD						
4"	4	8 – 3/32	4 – 5/32	3561	10.71	13.86
2.600	8	16 – 3/32	8 – 5/32	1031	17.92	
2.600	8	16 – 1/16	8 – 1/8	3571	17.92	
2.600	8	16 – 1/16	8 – 5/32	3581	17.92	
FORD – MERCURY						
3 5/16	8	8 – 5/64D				
		8 – 5/64	8 – 5/32	3591	17.92	
3 5/16	8	16 – 3/32	8 – 5/32	3601	17.92	24.22
3 5/16	8	16 – 3/32D	8 – 3/16	3611	17.92	24.22
3 5/16	8	16 – 1/16	8 – 5/32	3621	17.92	
3 5/16	8	16 – 3/32	16 – 5/32	3631	22.76	29.06
3 5/16	8	16 – 3/32D	16 – 3/16	3641	22.76	29.06
3 3/8	8	8 – 5/64D				
		8 – 5/64	8 – 5/32	3651	17.92	24.22
3 3/8	8	16 – 3/32	8 – 5/32	3661	17.92	24.22
3 3/8	8	16 – 3/32D	8 – 3/16	3671	17.92	24.22
3 3/8	8	16 – 1/16	8 – 5/32	3681	17.92	
3 3/8	8	16 – 3/32	16 – 5/32	3691	22.76	29.06
3 3/8	8	16 – 3/32D	16 – 3/16	3701	22.76	29.06
3 7/16	8	16 – 3/32	8 – 5/32	3711	17.92	24.22
3 1/2	8	16 – 3/32	8 – 5/32	3721	17.92	24.22
3 3/4	8	16 – 5/64D	8 – 3/16	1191	18.50	24.68
3 7/8	8	16 – 5/64D	8 – 3/16	2211	21.00	26.18
3 15/16	8	16 – 5/64D	8 – 3/16	11581	21.15	27.74
4"	8	16 – 5/64D	8 – 3/16	2711	20.90	26.80
G. M. C.						
3 15/16	6	6 – 3/32D				
		6 – 3/32	6 – 5/32	3731	16.06	20.79
4"	6	6 – 3/32D				
		6 – 3/32	6 – 5/32	3741	16.06	20.79
4 1/8	6	6 – 3/32D				
		6 – 3/32	6 – 3/16	3751	17.39	22.43
4 3/16	6	6 – 3/32D				
		6 – 3/32	6 – 3/16	3761	17.39	

CUSTOMIZING IS EASY WITH TAPKITS

TAPKIT

Contains 500 sq. in. specially treated "A" weight Fiberglas cloth, proper amount of TAPOX epoxy resin and hardener, milled Fiberglas fibres (for making filling paste), Stop-Sag powder (for use with resin to prevent run-off on vertical surfaces), mixing cups and spoons, brush, and complete illustrated instructions (illustrated). This is the kit for beginning "glassers" and for small repair jobs or a simple headlight frenching job.
NO. 3790 Tapkit C-10

TAPKIT

Contains 2,000 sq. in. of special "A" weight Fiberglas cloth, proper amount of TAPOX epoxy resin and hardener, milled Fiberglas fibres, Stop-Sag powder, mixing cups and spoons, brush, and complete illustrated instructions. This kit is adequate for hooding headlights or making an air scoop or installing fair-sized fins or patching several rust-outs and dents.
NO. 3791 Tapkit C-25

TAPKIT

Contains 18 sq. ft. of special "A" weight Fiberglas cloth, 8 lineal ft. 6" Fiberglas tape, two quarts of TAPOX epoxy resin with one pint hardener, 1 lb. milled Fiberglas fibres, one "C" kit of TAPOX Plastic Solder, mixing cups, spoons, and complete illustrated instructions. This is a "king size" kit containing ample materials for a complete fore-and-aft customizing job or for several professional repair jobs.
NO. 3792 Tapkit C-50

TAPOX RESIN

This is the amazing new plastic that bonds Fiberglas to any metal like a weld job. Like other resins used with Fiberglas, it comes in liquid form. When mixed with its hardener it hardens into a solid at normal room temperatures. Heat lamps will speed up the hardening process, of course. CAUTION: Only epoxy resin will bond Fiberglas to metal permanently. TAPOX is the finest epoxy resin that money can buy. This resin can also be used with Fiberglas to repair boats, furniture, etc. When you order TAPOX resin, you receive the proper amount of hardener without extra cost.
NO. 3808 8 oz. (avp. 1/2 pt) B-60 NO. 3810 32 oz. (avp. 1 qt) B-62
NO. 3809 16 oz. (avp. 1 pt) B-61 NO. 3811 128 oz. (avp. 1 gal) B-63

FIBERGLAS CLOTH

Specially treated woven glass fabric for use with TAPOX epoxy resin and all polyester type resins. Supplied in "A" weight -- correct for customizing and auto body repair work. Excellent for coating and repairing boat hulls, furniture, etc. Fiberglas cloth is used as a reinforcing material with resin. Two or three layers provide adequate strength for most applications.
NO. 3794 18" x 38" . . . . B-12 NO. 3793 36" x 38" B-11

STOP-SAG

This is a special powder which, when mixed into the resin-hardener mixture, will keep the "glass" in place on vertical surfaces while curing. Fine for bonding glass fins on rear fenders using Fiberglas tape. Use Stop-Sag in resin to keep the tape from slipping after it has been laid in place. Use Stop-Sag whenever you are "glassing" vertical surfaces.   NO. 3806 1 oz. B-51
NO. 3805 1/2 oz. . . . . . . B-50 NO. 3807 4 oz. B-52

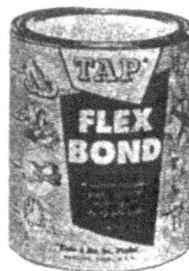

TAP FLEX-BOND SOLDER

- FASTER CONTROLLED CURE
- FEATHERS MORE SMOOTHLY
- MORE IMPACT STRENGTH
- BONDS BETTER
- FILES EASIER

Hardens tough as metal but also has controlled permanent flexibility. For this reason, FLEX-BOND can be used on large metal panels such as doors, hoods, trunk lids without fear of dislodging, cracking or "crazing" from impact or "oilcanning" of the metal underneath. Hammer on a FLEX-BOND patch. Chisel it. The metal underneath will fail before the patch! Furthermore, its inherent flexibility permits FLEX-BOND patches to expand and contract with metal during temperature changes - from Alaska to Death Valley. Because of its controlled flexibility, paint jobs stay new longer. No cracks in your repair job. . no cracks in your paint job.
NO. 3818 Pint - 1-1/2 lb. B-87
NO. 3819 Quart - 3 lb. B-88
(Above prices include both regular hardener and speed promoter)
NO. 3820 Hardener only - 1/2 oz. B-102
NO. 3821 Promoter only - 1/4 oz. B-111

TAP "500"

For the economy-minded, similar to Flex-Bond, however, for those who don't mind a little more work, TAP "500" is a new plastic solder reinforced with fiberglass and mineral fillers especially developed for repairing small holes and dents where impact strength and vibration is not a major factor. Can be used on larger areas or as a filler in conjunction with the regular glassing cloth on top. Hardener included with Tap "500".
NO. 3816 Pint - 1-1/2 # . . . . B-81 NO. 3815 Quart - 3# B-80

FIBERGLAS MATTE

This is a matting composed of short lengths of glass fibres, treated to insure quick wetting in resin. This material is used as a filler to quickly build up fender fins to final contour. Always top this material with at least one layer of Fiberglas cloth to provide sufficient strength to the job. Its prime use is as a "filler." For hoods, fins, etc.
NO. 3796 Fiberglas Matte 1-1/2 oz. 12" x 38" B-31
NO. 3795 Fiberglas Matte 1-1/2 oz. 36" x 38" B-30

TAPOX SOLDER

This material is supplied in kits consisting of two containers. One can contains a paste of TAPOX epoxy resin and metal granules, the other contains the hardener. Mix in equal parts, in quantities just sufficient for each job to avoid waste. This is pure "fix-it" magic for all craftsmen as well as customizers. Use it to fill small pits, cracks and holes on any surface. Excellent for reseating bolts and screws. When hardened, it can be sawed, drilled and tapped. Use for final shape and contour.
NO. 3812 Tapox Solder - Small ("A" Kit) . 1/8 pt. B-70
NO. 3813 Tapox Solder - Medium ("B" Kit) 1/2 pt. B-71
NO. 3814 Tapox Solder - Large ("C" Kit) . 1 pint B-72

MILLED FIBRES

These are chopped glass fibres supplied in bulk for mixing with resin to make a filling paste. For use in filling grooves and seams such as the groove between the headlight rim and fender in frenching jobs. This is a quick method for filling the groove around a cowl. Simply mix the fibres with the resin-hardener mixture and apply with a putty knife.
NO. 3802 Milled Fibres - 4 oz. B-40
NO. 3803 Milled Fibres - 8 oz. B-41
NO. 3804 Milled Fibres - 16 oz. B-42

FIBERGLAS TAPE

This is Fiberglas cloth fabric cut in various width strips and supplied in rolls for quick application. Excellent for dechroming jobs such as "bull-nosing" hoods.   NO. 3799 3" width . . B-22
NO. 3797 1-1/2" width B-20 NO. 3800 4" width . . B-23
NO. 3798 2" width . . B-21 NO. 3801 6" width . . B-24

W&H DuCoil — DUAL COIL IGNITIONS FOR ALL V-8 ENGINES

FULL VACUUM — CENTRIFUGAL ADVANCE

A complete new design of ignition, not a conversion. Features better than magneto output and performance in excess of 8000 RPM. Engineered to exactly replace stock distributor, requiring no alterations and may be used with all types of carburetion. Uses standard service parts available everywhere. DuCoil will equal, if not outperform any existing magneto or battery system under any condition yet costs less both to buy and maintain.

DuCoil "Competition" Dual Coil ignition featuring centrifugal advance only, has no vacuum control attached. Designed for engines where vacuum take-off is not available or needed (fuel injection, blowers, boats, dragsters, etc.). Can be used with any type of carburetion or induction system and may be ordered (no extra charge) with restricted advance curve ideally suited for supercharger installations #3670 .... $69.50
Thunderbird with tach drive #3671 .... $74.50

DuCoil "Street" Dual Coil centrifugal advance ignition with fast action full vacuum trimmer and precision BALL BEARING breaker plate. Vacuum control is utilized as additive factor to centrifugal curve providing high rate of initial advance. Designed primarily for use on engines where acceleration requirements cover the lower (500-2000) as well as the higher RPM ranges and low initial advance idling and starting are needed. Can be used with one or more carburetors when controlled (cut-off at idle) full (not venturi) vacuum is available at at least one carburetor. (Complete instructions included for altering Ford carburetors) #3672 .... $74.50
Thunderbird with tach drive. #3673 .... $79.50

DuPoint Single Coil Dual Point centrifugal advance ignition retaining all mechanical advance features of the DuCoil "Street" series. Ideally suited for stock and semi full race engines where Dual Coil electrical output and high RPM is not needed. Designed primarily for use on Ford products (49-56 venturi vacuum) when carburetion has been changed. #3674 .... $59.50
Thunderbird with tach drive. #3675 .... $64.50

STATE MAKE AND MODEL WHEN ORDERING

SPALDING FLAME THROWER IGNITIONS

Precision engineered, die-cast aluminum housings. Replaces stock ignitions with no special alterations. All units feature centrifugal advance with Full-Manifold Vacuum advance diaphragm. May be ordered for competition use only by leaving off the Vacuum diaphragh assembly. Comes with instructions for various Carburetor arrangements.

#3506 Spalding Flame Thrower Ignition, all 8 Cylinder V-8 engines or "T" Bird engines with tach drive   Ea. $108.00

1328, 1331

1329, 1330

HARMAN & COLLINS MAGNETO

Designed for V8 Ford & Mercury racing engines, this is a direct engine-connected, 8 pole rotor assembly and firing accuracy will be constant for life of unit. For 32-41 V8-85 and V8-60 . . . . Catalog No. 1329
For 42-48 models . . . . Catalog No. 1330

HARMAN & COLLINS DUAL COIL IGNITIONS

(No. 1400.) Positive insulation, standard advance, high speed cam, carbon track elimination, dual coil for longer more reliable service.
1942-48 . . . . Catalog No. 1328
1932-41 (use with distributor adapter plate) . Cat. No. 1331
(No. 1500) 1949-53 . . . . Catalog No. 1388

THE OBLIGATION OF

LEADERSHIP

Leadership in any industry is achieved and maintained only through the efforts of the leader to improve the quality, precision and efficiency of his products. Constant research, design and development plus a vast knowledge of the problems present are his tools.

Since its inception, Offenhauser Equipment Corporation has earned and enjoyed all of the benefits of its acknowledged leadership..... until today, the name of Offenhauser is synonymous with speed and power.

POWER IS NECESSARY

Yearly gain in horsepower ratings of all automobiles indicates the trend in power. This proves that power equipment is no longer an enthusiast's plaything, but is now a necessity.

Every user of special equipment creates his engine with preconceived performance standards. Too often, however, actual performance falls far short of his expectations. Research reveals that improper selection and application of equipment is the major cause of performance failures.

Selection of special equipment is based upon the uses to which the engine will be subjected ... road driving or competition ... altitude and humidity of his general locale. These and many other generally ignored factors are a part of the careful engineering of all speed equipment.

Nor is the addition of more speed equipment necessarily the answer. While the addition of special equipment may increase the performance slightly, it may actually retard the efficiency of other special equipment. Each part must coordinate perfectly... improve the efficiency of each *other* part. The result is perfect coordination .... a perfectly engineered engine that *can* therefore fulfill every expectation.

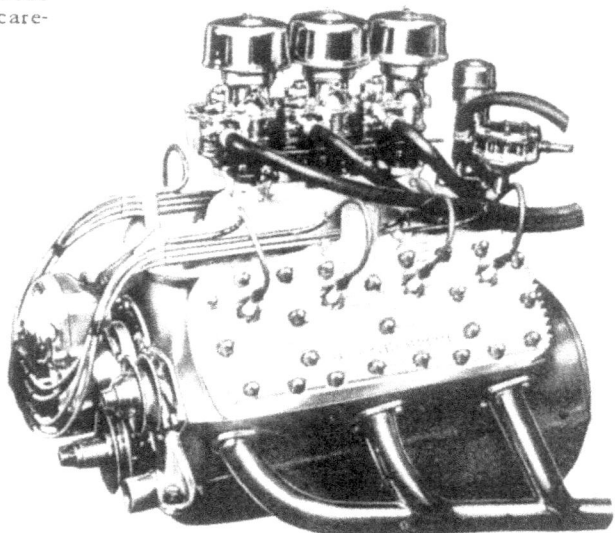

Offenhauser
EQUIPMENT CORP.

CApitol 5-1307
5156 Alhambra Avenue
Los Angeles 32, California

www.ingramcontent.com/pod-product-compliance
Lightning Source LLC
Chambersburg PA
CBHW082110210326